SpringerBriefs in Physics

T0202844

For further volumes:
http://www.springer.com/series/8902

Helge S. Kragh · James M. Overduin

The Weight of the Vacuum

A Scientific History of Dark Energy

Helge S. Kragh
Centre for Science Studies
Aarhus University
Aarhus
Denmark

James M. Overduin
Department of Physics, Astronomy
  and Geosciences
Towson University
Towson, MD
USA

ISSN 2191-5423
ISBN 978-3-642-55089-8
DOI 10.1007/978-3-642-55090-4
Springer Heidelberg New York Dordrecht London

ISSN 2191-5431   (electronic)
ISBN 978-3-642-55090-4   (eBook)

Library of Congress Control Number: 2014938218

Printed on acid-free paper

Springer is part of Springer Science+Business Media (www.springer.com)

Preface

Dark energy is a catch-all term for the energy of empty space. It is "dark," not just in the sense that it does not interact with electromagnetic radiation, but in the deeper sense that its nature and composition are essentially unknown. Dark energy is known to dominate the dynamics of the universe on large scales, and to oppose the natural tendency of the cosmos to collapse under the weight of its own contents. In fact, under the influence of dark energy, the universe has entered a period of "late-term inflation" in which the expansion of space has started to *accelerate*, and will never stop accelerating, world without end.

The idea of an energetic vacuum has many historical antecedents, from the *pneuma* of the Stoic philosophers to the luminiferous ether of the late nineteenth century. It gained particular urgency with the advent of quantum theory, when it was realized that all fields in nature are endowed with irreducible "zero-point energies," even in their vacuum states. As we now recognize, the sum of these zero-point energies exceeds by some 120 orders of magnitude the actual energy density of the universe, a mismatch so severe that it has rightly been called by Steven Weinberg the "one veritable crisis" left in physics. The full sting of this crisis is not felt within quantum field theories themselves, where such energies can be mathematically absorbed by a procedure known as renormalization. But it becomes unavoidable when such theories are confronted with the one field in nature that has not yet been successfully quantized: gravity. For according to general relativity, *all forms of energy must gravitate*. Thus the crisis is, in essence, a confrontation between the quantum field-theoretic approach to nature that we have inherited from Democritus, and the geometric picture of gravity as curved spacetime that can be traced from Einstein back to Pythagoras. And its resolution is likely to be bound up with the dream of physicists and philosophers through the ages to unify all the laws of nature in one single, all-encompassing "theory of everything."

We have written this book with two goals in mind. The first is to look back over the theoretical precursors to modern dark energy, and attempt to put the subject in some historical perspective. Such a perspective may have its uses but cannot, of course, lead on its own to the kind of radical new thinking that will be required to solve the underlying issue. The story of the ether is instructive here. Physics at the end of the nineteenth century was gripped by a similar crisis when it was realized that light did not obey the same laws as everything else. Then too, theorists looked back over history and attempted to extend or modify existing concepts in order to accommodate this new mystery. But a successful resolution—and with it a unified theory of electromagnetism and the rest of nineteenth-century physics—came only when Einstein proved able to jettison a belief so axiomatic that most did not even realize they held it: the independence of space and time. It is likely that dark energy—and with it the unification of gravity and the standard model of particle physics—will similarly be understood by looking, not backward, but *sideways*, at some unstated and apparently unrelated assumption so fundamental that it has previously gone unquestioned.

Our second goal is to record some of the observational hints that preceded the celebrated discovery of dark energy by cosmologists in 1998. It was then that the theoretical crisis assumed its true proportions, for this discovery removed any hope that the standard model was merely incomplete, and that some new under-lying symmetry would be found which would ensure an exact cancellation of all the zero-point contributions. In fact, the energy density of the cosmic vacuum is *not* zero, but apparently exists at the preposterously fine-tuned level of about 10^{-120} times that which is expected on the basis of standard (and exceedingly well-tested) quantum field theory. The fact that such a result became widely accepted in considerably less time than the prior proposal of dark matter (not to mention the original discovery of cosmic expansion itself) testifies to the strength of the observational case built by the supernova discovery teams. In such cases it can become possible to see the progress of science as illuminated mainly by the blaze of searchlights. We hope to show how it has also progressed by the aid of candles, sometimes shining in the wrong places altogether, whose light never-theless grows in fits and starts until it makes the searchlights possible.

In writing this book we have benefitted from the insights of many people. For discussions over the years we thank Ron Adler, Sean Carroll, Arthur Chernin, Fred Cooperstock, Francis Everitt, Hans-Jörg Fahr, Masataka Fukugita, Takao Fukui, Eric Gliner, Leopold Halpern, Joohan Lee, Kei-Ichi Maeda, Bahram Mashhoon, Simon Overduin, Wolf Priester, Bharat Ratra, Varun Sahni, Alex Silbergleit, Bob Wagoner, Paul Wesson, and Richard Woodard among others. The errors that remain are, of course, our own.

March 2014   Helge S. Kragh
                                                                  James M. Overduin

Contents

Chapter 1
Early Ideas of Space and Vacuum

Abstract Aristotle's idea of a heavenly element (*quinta essentia*) and his arguments against a void were generally accepted in the ancient world and through the middle ages. However, other ideas circulated as well, such as the active *pneuma* postulated by Stoic philosophers. Following experiments with barometers and air-pumps in the seventeenth century, the hypothesis of a natural *horror vacui* lost credibility. Empty space had been discovered in the laboratory—but was the space without air found by Robert Boyle and others really empty?

Keywords Vacuum · Plenum · Aristotle · Pneuma · Horror vacui

The concept of empty space—void, vacuum, or sometimes nothingness—has a fascinating history that goes back to the ancient Greeks and can be followed all the way to the present. The brief version of the story is this. For nearly two millennia a true vacuum was thought to be an impossibility, or at least something which does not exist in nature; then, in the seventeenth century the new experimental physics proved that a vacuum can in fact be produced and made a study of inquiry; but later generations of physicists realized that this vacuum is a far cry from the absolute empty space discussed by the ancients. In any case, physicists' vacuum or so-called empty space is entirely different from the metaphysical concept of nothingness.[1]

According to some Presocratic thinkers, notably the atomistic school associated with Leucippus and Democritus, a vacuum was not only possible but also necessary. The entire natural philosophy of the atomists rested on the postulate that the indivisible elements of matter, called atoms, move in an absolute vacuum. This is all there is, atoms and empty space, and the latter is no less real than the former. According to Simplicius, a Greek commentator from the early sixth century AD, "supposing the substance of the atom to be compact and full, he [Leucippus] said it is 'being' and that it moves in the void, which he called 'non-being' and which he declares is no less than what is" (McKirahan 1994, p. 306). Archytas, a Pythagorean

---

[1] The history and physical meaning of empty space is the subject of several books and articles. Popular accounts include Genz (1999), Barrow (2001) and Close (2009).

H. S. Kragh and J. M. Overduin, *The Weight of the Vacuum*, SpringerBriefs in Physics, DOI: 10.1007/978-3-642-55090-4_1, © The Author(s) 2014

and contemporary of Plato, asserted that "Since everything which is moved is moved into a certain place, it is plain that the place where the thing moving or being moved shall be, must exist first" (Jammer 1954, p. 10). However, these were views of which Aristotle, the most influential thinker ever in the history of science and ideas, would have nothing.

Aristotle's cosmos was a plenum, and his basic concepts of space and place ruled out a vacuum, whether on a small or large scale. In *Physics* and other of his works, he launched an impressive number of objections against void space, which he defined as a place devoid of body, but capable of receiving it. Almost all of his arguments were thought experiments or of the *reductio ad absurdum* type where the assumption of a vacuum was shown to lead to absurd conclusions. For example, by its very nature empty space is completely homogeneous and for this reason admits no difference. Every part of it is identical to every other part, and so there can be no spatial orientation in a void and no measurement of distances either. What Aristotle called natural motion was, insofar as it happens on the Earth and not in the heavens, either straight up or straight down. Whereas the elements fire and air naturally move upward, water and earth move downward. But in empty space there is no "up" and "down," and so a stone would not know how to move. According to ancient thought bodies move because there is a reason for it, but in a vacuum there can be no reason for moving in a particular direction rather than in some other direction.

Not only would there be no reason or purpose for a particular motion in a void, there might be no motion at all. Another of Aristotle's arguments related to falling bodies, the velocity v of which he claimed was governed by the weight (or size) of the body and the resistance of the medium in which it moves. Somewhat anachronistically Aristotle's "law of motion" may be written

$$v = k\frac{F}{R},\qquad\qquad(1.1)$$

where k is a constant and R the friction between the body and the surrounding medium. F is the motive force, assumed to be proportional to the weight or *gravitas* of the body. Thus, a body falling freely in empty space would, irrespective of its weight, size and material, move with infinite velocity. Clearly, instantaneous natural motion must be ruled out as impossible. Even if motion in a void were finite, bodies of different weight would fall with equal velocities, contrary to Aristotle's principles of natural philosophy. Because there is no medium in a void, no reason can be given why one body should move with a greater velocity than another.

Aristotle's ether, as he discussed it most fully in his influential work *De Caelo*, was a divine substance or fifth element, known by his medieval successors as *quinta essentia*. It made up everything in the heavens—the part of the universe above the Moon—and was, unlike the matter of the sublunar world, pure, ungenerated, incorruptible, and transparent. While the vast expanse between the celestial bodies appears to be empty, in reality it is filled up with the subtle matter or quintessence. According to some modern commentators, Aristotle's ether is close in spirit to the dark cosmological energy (Krauss 2000, p. 222; Decaen 2004). However, there is the obvious difference that whereas dark energy is everywhere in space, the ether of the Aristotelians was strictly limited to the celestial regions. And, of course, this is not the only difference.

Whereas Aristotle's space and quintessential ether were essentially passive qualities, other Greek thinkers proposed that all of space was filled with a continuous active principle. The Stoic school of philosophy agreed that there was no empty space, and their space-filling *pneuma* was a means to avoid it. However, pneuma was a vital elastic substance that could change in both time and place. Stoic philosophers conceived the plenum as a mixture of the elements fire and air, the mixture being associated with a dynamical function embracing all natural phenomena. The pneuma filled the whole universe, both the space between bodies and the bodies themselves. According to Chrysippos, a philosopher of the third century BC, "the whole of nature is united by the pneuma which permeates it and by which the world is kept together and is made coherent and interconnected" (Sambursky 1963, p. 135). This universal binding force has been likened to the physical field of later science and also to the kind of ether that emerged at the time of Newton.[2] On the other hand, Aristotle's ether or fifth element was restricted to the heavens, and in this respect it differed from *pneuma*.

There were similarities as well as dissimilarities between the Aristotelians and the Stoics. The latter followed Aristotle in conceiving the physical cosmos as a finite sphere without any vacua whatever. They also adopted Aristotle's definition of vacuum as a space that could contain body, but in fact did not. Nonetheless, they supposed the most un-Aristotelian notion of an immense void that was not only extra-cosmic but also infinitely extended. This extra-cosmic void was a three-dimensional container in which the finite spherical cosmos was embedded (Fig. 1.1). They thought that the two parts of the universe at large were strictly separate: the infinite void space had no properties of its own and thus there could be no physical interaction between it and the physical cosmos. Matter could not dissipate into the void, for "the material world preserves itself by an immense force, alternately contracting and expanding into the void following its physical transformations, at one time consumed by fire, at another beginning again the creation of the cosmos" (Sambursky 1963, p. 203). The Stoics' universe was cyclic, which was another non-Aristotelian feature.

By the mid-thirteenth century Aristotle's entire work had been translated into Latin and enthusiastically received by most medieval scholars. Whereas they, as pious Christians, were bound to deny Aristotle's conclusion of an eternal, uncreated world, there were no serious obstacles to other of the principles of "the philosopher." Thus, in the high middle ages it was generally granted that "nature abhors a vacuum" (Grant 1981). Although rarely questioned, the doctrine of *horror vacui* was endlessly debated by the learned philosophers, who often related it to God's omnipotence.

---

[2] Of particular interest is the Stoic contrast between the outward-directed tension associated with *pneuma* and the innate tendency of the cosmos to move "toward its own center" (Sambursky 1959), which anticipate in a metaphorical way the contrasting dynamical effects of dark energy and gravitating matter on large scales in modern cosmology.

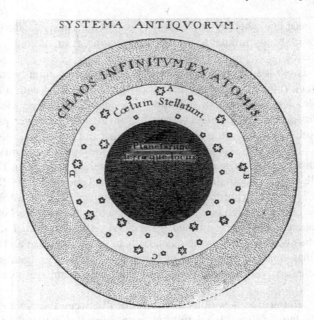

Fig. 1.1 This depiction of the universe, from a book published in England in 1675, includes elements of both the Stoic pneuma and the atomists' infinite chaos

To be sure, a void does not exist naturally, nor can it be produced artificially—but could God have created one if he so wished?

The schoolmen at the new universities in Oxford, Paris and elsewhere also discussed at length whether or not the corporeal world was surrounded by an immense void space, as assumed by the Stoics. The distinguished Paris philosopher Jean Buridan represented the majority view when he wrote: "An infinite space existing supernaturally beyond the heavens or outside this world ought not to be assumed" (Grant 1994, p. 170). Yet he qualified his statement by adding that beyond this world God could create whatever space and corporeal substances it pleased him to create. Although medieval philosophers generally followed Aristotle's denial of an extra-cosmic void, several of them considered the idea of an infinite "imaginary space" that in some sense encompassed the physical universe. This space was a kind of vacuum, but very different from the one discussed by the ancients. It was dimensionless and unable to contain matter. In effect, it was a theological construct, God's infinite abode and an abstract expression for his immensity. Among those who subscribed to such a view was Thomas Bradwardine, an eminent Oxford mathematician and natural philosopher. Although he followed many of Aristotle's arguments against a void, Bradwardine nevertheless maintained that the world was surrounded by an extra-cosmic vacuum, void of any body and of everything other than God.

The popularity of Aristotelianism caused some concern among leading theologians who feared that some of the views held by radical philosophers might undermine established church dogmas. As the result of the growing controversy between faith and reason, in 1277 the Bishop of Paris, Etienne Tempier, issued a list of no less than 219 propositions that were declared false and heretical. Among the heresies was "That God could not move the heavens with rectilinear motion; and the reason is that a vacuum would remain." Surely, the omnipotent deity was not bound to follow Aristotle's rules. This and other of the condemnations did not silence the philosophers, but it is of interest because it indicates that the Aristotelian *horror vacui* did not enjoy unrestricted ecclesiastical support. Rather on the contrary, for it was now confirmed that God could create a vacuum and consequently it became admissible to discuss various kinds of vacua at least hypothetically. On the other hand, it remained largely undisputed that vacua did not actually exist in nature.

This consensus view was questioned during the scientific revolution in the seventeenth century. Galileo argued that although nature in a certain sense abhorred a vacuum, it was not in an absolute sense. A state of vacuum, in the sense of a portion of space devoid of air, might be produced experimentally. Evangelista Torricelli, Galileo's last assistant, greatly developed his master's suggestion and in 1644 succeeded in evacuating the air from the space above a mercury column. This he considered a kind of vacuum, if an incomplete and unnatural one. When Blaise Pascal in France learned about Torricelli's work, he improved it in a series of brilliant experiments and provided the observed phenomena with a clearer and better argued explanation. According to Pascal, all the phenomena that traditionally had been ascribed to nature's *horror vacui* were in reality effects of the weight and pressure of the surrounding air. "Nature does nothing to avoid the vacuum," he wrote. "Rather, the weight of the air masses is the true reason for all these phenomena which we have been ascribing to an imaginary cause" (Genz 1999, p. 121). Pascal even weighed the vacuum, meaning that he showed that it had no measurable weight compared to air. By 1660 the hypothesis of *horror vacui* was obsolete, if still alive. New Aristotelian versions of it continued to be proposed, but they were generally considered inferior to the new explanations based on the pressure of air.

The first air-pumps, devices to evacuate air from closed containers, followed in the wake of the discoveries of Torricelli and Pascal. At about 1650 Otto von Guericke in Magdeburg, Germany, invented his first version of a pneumatic pump, which some years later was turned into an improved machine by Robert Boyle in England (Boyle's pumps were actually constructed by his assistant Robert Hooke, an ingenious experimenter). Guericke and Boyle not only demonstrated the elasticity of air—or what Boyle called the "spring of air"—they also investigated some of the properties of the produced vacuum. For example, whereas sound could not propagate through a vacuum, it did not hinder the propagation of either magnetism or light. Both of the inventor-philosophers speculated about the nature of the evacuated space, discussing whether it was a true vacuum or not. As to speculation and imagination, Guericke far outstripped Boyle. He claimed that everything created had its origin and place in what he called "nothing" and which he conceived as an infinite extra-cosmic void indistinguishable from God. While the kind of vacuum produced by the air-pump was a three-dimensional empty space, Guericke's true nothingness had no dimensions. Nonetheless, he claimed it was real.

Contrary to Boyle, Guericke considered his vacuum to be of deep cosmological significance and part of his argument for an infinite universe populated by an infinity of stellar bodies. In his main work with the abbreviated title *Nova Experimenta Magdeburgica* he explained that, far from being absolutely void and physically impotent, the nothing of infinite space was an active and powerful entity. It was real and imaginary at the same time. In what has been called an "ode to nothing" he expressed the vacuum-nothingness in lyrical language (Grant 1981, p. 216):

> Nothing contains all things. It is more precious than gold, without beginning and end, more joyous than the perception of bountiful light, more noble than the blood of kings, comparable to the heavens, higher than the stars, more powerful than a stroke of lightning, perfect and blessed in every way. Nothing always inspires. ... Nothing is outside the world. Nothing is everywhere.

When it came to the question of whether or not the air-pump produced a true vacuum, Boyle was much more restrained than his fellow-physicist in Magdeburg. He refused to speculate about the relationship between the evacuated space and the traditional meaning of vacuum, which to him was a "metaphysical" concept. Instead he adopted an operationalist notion of the vacuum, simply identifying it with the space from which all air had been removed. Boyle realized that even the most completely evacuated space might not be absolutely empty, since there was always the possibility of the presence of some undetectable ethereal substance. But for him, as an experimental philosopher, it was a problem that could be safely ignored and left to the metaphysicians.

References

Barrow, J.: The Book of Nothing. Vintage, London (2001)
Close, F.: Nothing: A Very Short Introduction. Oxford University Press, Oxford (2009)
Decaen, C.A.: Aristotle's ether and contemporary science. The Thomist **68**, 375–429 (2004)
Genz, H.: Nothingness: The Science of Empty Space. Basic Books, New York (1999)
Grant, E.: Much Ado About Nothing: Theories of Space and Vacuum from the Middle Ages to the Scientific Revolution. Cambridge University Press, Cambridge (1981)
Grant, E.: Planets, Stars, & Orbs: The Medieval Cosmos, 1200–1687. Cambridge University Press, Cambridge (1994)
Jammer, M.: Concepts of Space: The History of Theories of Space in Physics. Harvard University Press, Cambridge (Mass) (1954)
Krauss, L.: Quintessence: The Mystery of Missing Mass in the Universe. Basic Books, New York (2000)
McKirahan, R.: Philosophy Before Socrates. Hackett Publishing, Indianapolis (1994)
Sambursky, S.: Physics of the Stoics. Princeton University Press, Princeton (1959)
Sambursky, S.: The Physical World of the Greeks. Routledge, London (1963)

Chapter 2
The Active Ether

Abstract During most of the nineteenth century the existence of an all-encompassing ether as a substitute for a true void was taken for granted. According to the successful electromagnetic ether theory based upon Maxwellian electrodynamics, the ether was endowed with energy and hence very different from nothingness. Oliver Lodge thought that its energy density was enormous, about 10^{33} erg cm^{-3}. The active ether of the fin-de-siècle period was in some respects surprisingly similar to the later quantum vacuum, yet it was based fully on classical physics.

Keywords Ether · Electromagnetism · Vortex theory · Oliver Lodge

In spite of the opinion of Boyle, many later natural philosophers assumed the existence of a rare and penetrating "subtle matter" that was present also in a void. Although some physicists accepted the vacuum as a passive nothingness, this was not the generally held view. The period from Newton to Maxwell saw a bewildering variety of ethers which in many cases were introduced for specific purposes, such as explaining electricity, magnetism, light, gravitation, nervous impulses, and chemical action (Cantor and Hodge 1981). In an article on the ether for the *Encyclopaedia Britannica*, Maxwell (1965, p. 763) noted that "To those who maintained the existence of a plenum as a philosophical principle, nature's abhorrence of a vacuum was a sufficient reason for imagining an all-surrounding æther." The fact that some of these phenomena were easily transmitted in empty space indicated that their associated ethers were part of even a perfect vacuum. Maxwell continued: "Æthers were invented for the planets to swim in, to constitute electric atmospheres and magnetic effluvia, to convey sensations from one part of our bodies to another, and so on, till a space had been filled three or four times with æthers."

As he implied here, the many hypothetical ethers were *ad hoc* and therefore unsatisfactory from a methodological point of view. Yet Maxwell was himself not only a believer in the ether; his field theory of electromagnetism was instrumental in introducing a new and unified kind of ether, one which was electrodynamical rather than mechanical in nature. Since the early decades of the nineteenth century the ether had become increasingly associated with optics and seen as the medium in

H. S. Kragh and J. M. Overduin, *The Weight of the Vacuum*, SpringerBriefs in Physics, DOI: 10.1007/978-3-642-55090-4_2, © The Author(s) 2014

which light propagated. Following the pioneering work of Thomas Young in England and Augustin Fresnel in France, by the 1820s the corpuscular theory of light was abandoned and replaced by a theory of transverse waves. The new "luminiferous" ether pervaded the universe and, according to most physicists, behaved like an elastic solid that—strangely—did not interact with other matter. Although it had the form of a solid, and was sometimes likened to steel, the planets and comets passed through it without noticing any resistance. Strange indeed!

In a postscript to an 1864 article titled "A dynamical theory of the electromagnetic field," Maxwell (1864), Maxwell even wondered whether the electromagnetic ether might in some way be responsible for the force of gravitation. His idea seemed to be that by "screening" each other from a portion of this energetic medium, massive bodies might be driven together as if they were accelerating under the influence of a new force. After noting that the energy of the ether is everywhere positive, and attempting to obtain a lower limit on its density, he concluded:

> The assumption, therefore, that gravitation arises from the action of the surrounding medium... leads to the conclusion that every part of this medium possesses, when undisturbed, an enormous intrinsic energy, and that the presence of dense bodies influences the medium so as to diminish this energy wherever there is a resultant attraction.
>
> As I am unable to understand in what way a medium can possess such properties, I cannot go any further in this direction in searching for the cause of gravitation.

This is a remarkable anticipation, not only of the zero-point energy of quantum theory some 50 years later (Chap. 4), but also of attempts by some modern unified-field theorists to account for gravitational interactions as a Casimir-type byproduct of electromagnetic zero-point fields.[1]

Much ingenuity and mathematical effort was put into the construction of ether theories by theoretical physicists such as George Green, Gabriel Stokes, and James MacCullagh (Schaffner 1972). Ethers of the elastic-solid type were developed until the 1880s, often without connection to phenomena of electrodynamics. With the gradual acceptance of Maxwell's electromagnetic theory of light, as fully expounded in his monumental *Treatise on Electricity and Magnetism* from 1873, the ether came to be seen as inextricably associated with electromagnetism. Through the works of Hendrik A. Lorentz, Joseph Larmor, Oliver Heaviside, Max Abraham and many other theorists, Maxwell's field theory was developed into a sophisticated theory of the electromagnetic ether. Characteristically, when the German physicist Paul Drude in 1894 wrote an advanced textbook of Maxwellian electrodynamics, he chose to entitle it *Physik des Aethers* (Physics of the Ether). Whatever its precise nature, the ether was considered indispensable. Another German physicist and specialist in electrodynamics, August Föppl, suggested that space without ether would be a contradiction in terms, like a forest without trees.

Late nineteenth-century physics consisted of the physics of matter and the physics of the electromagnetic ether. To avoid the unwanted dualism, the trend was to identify matter with ether, rather than the other way around. In the early years of the twentieth

---

[1] The original idea for such a theory is usually credited to in Sakharov (1968). In its most compelling form it has been developed under the name "stochastic electrodynamics" by Haisch et al. (1994).

century it became common to regard the new electron as a concentration of or singularity in the ether; or, what was about the same, electromagnetic fields. This was a basic assumption of the so-called electromagnetic world picture that held a strong position in theoretical physics. But the views concerning space, ether, and vacuum differed. In a lecture of 1909 at Columbia University, Max Planck said: "In place of the so-called free ether there is now substituted the absolute vacuum,.... I believe it follows as a consequence that no physical properties can be consistently ascribed to the absolute vacuum" (Planck 1915, p. 119). He regarded the speed of light not as a property of the vacuum, but a property of its electromagnetic energy: "Where there is no energy there can be no velocity of propagation." Two years later Planck would initiate a development that led to the modern view of a true vacuum endowed with physical properties.

It is possible to trace the concept of dark energy far back in time, say to the days of Newton (Calder and Lahav 2008) or even to the *pneuma* of the Stoic philosophers. However, if one wants to point to pre-quantum and pre-relativity analogies to dark energy, a more sensible arena might be the ethereal world view of the late nineteenth century. The general idea that cosmic as well as terrestrial space is permeated by an unusual form of hidden energy—a dark energy of some sort—was popular during the Victorian era, where space was often identified with the ether. The generally accepted ethereal medium existed in many forms, some of them assuming the ether to be imponderable while others assumed that it was quasi-material and only differed in degree from ordinary gaseous matter in a highly rarefied state.

The ether was sometimes thought of as a very tenuous, primordial gas, perhaps consisting of ether atoms of the incredibly small mass 10^{-45} g. On the other hand, according to the popular vortex theory, which was cultivated by British physicists in particular, the discreteness of matter (atoms) was epiphenomenal, derived from stable dynamic configurations of a perfect fluid. This all-pervading fluid was usually identified with the continuous and frictionless ether. The highly ambitious vortex theory invented by William Thomson, the later Lord Kelvin, was not only a theory of atoms, it was a universal theory of ether (or space) and matter, indeed of everything (Kragh 2002).

The point is that by the turn of the nineteenth century few physicists thought of "empty space" as really empty. Rather it was filled with an active ethereal medium. This ether was widely seen as "a perfectly continuous, subtle, incompressible substance pervading all space and penetrating between the molecules of all ordinary matter, which are embedded in it" (Lodge 1883, p. 305). Lorentz and other physicists in the early twentieth century often spoke of the ether as equivalent to a vacuum, but it was a vacuum that was far from nothingness. Although Lorentz was careful to separate ether and matter, his ether was "the seat of an electromagnetic field with its energy and its vibrations,... [and] endowed with a certain degree of substantiality" (Lorentz 1909, p. 230). On the other hand, the popular belief in a dynamically active ether was rarely considered in astronomical or cosmological contexts.

Among the firm believers in the ether as a storehouse of potential energy was the English physicist Oliver Lodge (Fig. 2.1), a devoted follower of Maxwell who has been called a "remote ancestor" of the modern quantum vacuum. Peter Rowlands, a

Fig. 2.1 The English physicist Oliver Lodge (1851–1940), a pioneer in radio telegraphy and a life-long advocate of the electromagnetic ether

biographer of Lodge, comments: "The infinite energy density of the zero-point vacuum field fluctuations is almost indistinguishable from the infinite elasticity of the universal ethereal medium" (Rowlands 1990, p. 285). Lodge was indeed an enthusiastic protagonist of the active Victorian ether, which he considered incompressible and a reservoir of an immense amount of energy. This energy was not directly testable, but it could be calculated. In one such calculation, dating from 1907, he estimated the minimum etherial energy density to be "something like ten-thousand-million times that of platinum." In the words of Lodge (1907, p. 493):

> The intrinsic constitutional kinetic energy of the æther, which confers upon it its properties and enables it to transmit waves, is thus comparable with 10^{33} ergs per c.c.; or say 100 foot-lbs. per atomic volume. This is equivalent to saying that 3×10^{17} kilowatt-hours, or the total output of a million-kilowatt power station for thirty million years, exists permanently, and at present inaccessibly, in every cubic millimetre of space.

The energy density of Lodge's ether, if transformed to a mass density by means of $E = mc^2$, corresponded to about 10,000 tons cm^{-3}. In a later chapter on the possible granular structure of the ether, he repeated the estimate of 10^{30} to 10^{33} erg cm^{-3}, adding that "the ether may quite well contain a linear dimension of the order 10^{-30}

to 10^{-33} centim." (Lodge 1920, p. 171). Although not more than a curiosity, it is worth pointing out that the linear dimension of Lodge's ether happened to be in the same range as what would later become known as the Planck length,

$$\sqrt{\frac{Gh}{c^3}} = 4 \times 10^{-33}\,\text{cm}. \tag{2.1}$$

A length scale of the same order appears in modern unified theories of the fundamental interactions, such as string theory.

As another example, perhaps an even more dubious ancestor, consider the French psychologist and amateur physicist Gustave LeBon, the discoverer of the illusory "black light" and author of the best-selling *The Evolution of Matter*. In this time-typical and hugely popular book, LeBon (1905, pp. 313–315) pictured electrons and other charged particles as intermediates between ordinary matter and the ether. His cosmic scenario started with "a shapeless cloud of ether" which somehow was organized into the form of energy-rich atomic particles. However, these would be radioactive and slowly release their energy. They were "the last stage but one of the disappearance of matter," the last stage being represented by "the vibrations of the ether." Matter formed by electric particles would eventually radiate away all their stored energy and return to "the primitive ether whence they came... [and which] represents the final nirvana to which all things return after a more or less ephemeral existence."

The analogy between some forms of the classical ether and the presently discussed vacuum energy has not been lost on modern physicists, who sometimes refer to the resurrection or "transmogrification" of the ether (Sciama 1978). Even before the discovery of the cosmic microwave background, Robert Dicke at Princeton University wrote: "One suspects that, with empty space having so many properties, all that had been accomplished in destroying the ether was a semantic trick. The ether had been renamed the vacuum" (Dicke 1959). According to Paul Davies (1982, p. 582), late-nineteenth century physicists "would surely have been gratified to learn that in its modern quantum form, the ether has materialised at last." Similarly referring to the role of virtual particles in modern quantum field theory, Nobel laureate Frank Wilczek suggests that the ether—"renamed and only thinly disguised"—plays a most important role in physics (Wilczek 1999; see also Barone 2004). Tempting as it may be to consider the classical ether as an anticipation of modern vacuum energy, we need to emphasize that it is at most a crude analogy and that the historical connection between the two concepts is largely a reconstruction with little support in actual history. Vacuum energy is a quantum phenomenon and to find its historical origin we need to look at the early development of quantum theory.

References

Barone, M.: The vacuum as ether in the last century. Found. Phys. **34**, 1973–1982 (2004)

Calder, L., Lahav, O.: Dark energy: Back to Newton? Astron. Geophys. **49**, 1.13–1.18 (2008)

Cantor, G.N., Hodge, M.J.S. (eds.): Conceptions of Ether: Studies in the History of Ether Theories. Cambridge University Press, Cambridge (1981)

Davies, P.: Something for nothing. New Sci. **94**, 580–582 (1982)

Dicke, R.H.: Gravitation—an enigma. Am. Sci. **47**, 25–40 (1959)

Haisch, B., Rueda, A., Puthoff, H.E.: Inertia as a zero-point-field Lorentz force. Phys. Rev. A **49**, 678–694 (1994)

Kragh, H.: The vortex atom: a Victorian theory of everything. Centaurus **44**, 32–115 (2002)

LeBon, G.: The Evolution of Matter. Charles Scribner's Sons, New York (1905)

Lodge, O.: The ether and its functions. Nature **34**(304–306), 328–330 (1883)

Lodge, O.: The density of the æther. Phil. Mag. **13**, 488–506 (1907)

Lodge, O.: Note on a possible structure for the ether. Phil. Mag. **39**, 170–174 (1920)

Lorentz, H.A.: The Theory of Electrons. Teubner, Leipzig (1909)

Maxwell, J.C.: A dynamical theory of the electromagnetic field. Trans. R. Soc. (Lond.) CLV, 570–571 (8 Dec 1864)

Maxwell, J.C.: The Scientific Chapters of James Clerk Maxwell. Niven, W.D. (ed.) Dover Publications, New York (1965)

Planck, M.: Eight Lectures on Theoretical Physics. Columbia University Press, New York (1915)

Rowlands, P.: Oliver Lodge and the Liverpool Physical Society. Liverpool University Press, Liverpool (1990)

Sakharov, A.D.: Vacuum quantum fluctuations in curved space and the theory of gravitation. Sov. Phys. Doklady **12**, 1040 (1968)

Schaffner, K.F.: Nineteenth-Century Aether Theories. Pegamon Press, Oxford (1972)

Sciama, D.W.: The ether transmogrified. New Sci. **77**, 298–300 (1978)

Wilczek, F.: The persistence of the ether. Phys. Today **52**(January), 11–12 (1999)

Chapter 3
Planck's Second Quantum Theory

Abstract The notion of a zero-point energy is a result of quantum theory and has no proper counterpart in classical physics. It was introduced by Planck in 1911, more than a decade before the emergence of modern quantum mechanics. Planck's so-called second quantum theory, on which the zero-point energy was based, was discussed for a brief period of time, but by 1920 at the latest it was abandoned by most physicists. On the other hand, although Planck's theory was dismissed, the idea of a zero-point energy lived on. No one could tell whether it was more than just an idea.

Keywords Quantum theory · Max Planck · Zero-point energy · Heat radiation

According to Max Planck's epoch-making theory of 1900 the energy of an oscillator of frequency v was given by $E = hv$, where h was a new constant of nature with the dimension of an action (Planck's original value was $h = 6.55 \times 10^{-27}$ erg s). If a black body emitted or absorbed energy it would do so discontinuously, the energy changing in multiples of the quantity hv:

$$E = nhv, \qquad n = 0, 1, 2, \ldots \tag{3.1}$$

However, Planck was uncertain about the nature of the "energy elements" or quanta and tended to regard as formal rather than physically real. During the years 1911–1913 he reformulated his theory in such a way that it became more understandable from a classical point of view, and in this process he introduced a major conceptual novelty, the zero-point energy. He first introduced his new radiation hypothesis or "second theory" in an address to the German Physical Society in February 1911, and he subsequently developed it in several papers and lectures, including his report on heat radiation delivered to the first Solvay conference in Brussels later that year (Planck 1911, 1912, 1913, pp. 132–145). The new theory became more widely known

H. S. Kragh and J. M. Overduin, *The Weight of the Vacuum*, SpringerBriefs in Physics, DOI: 10.1007/978-3-642-55090-4_3, © The Author(s) 2014

Fig. 3.1 Planck (*right*) in a discussion with Niels Bohr in Copenhagen in 1930

from the exposition which appeared in the second edition of Planck's *Theorie der Wärmestrahlung* (Theory of Heat Radiation) published in early 1913.[1]

Whereas Planck (Fig. 3.1) in his original theory of 1900 had treated emission and absorption of radiation symmetrically, in his second theory—at the time generally known as the "theory of quantum emission"—he assumed that only the emission of radiation occurred in discrete energy quanta. The emission of these quanta would be governed by a probabilistic law of electrodynamics. Absorption, on the other hand, was supposed to occur in accordance with classical theory, that is, continuously. This feature appealed to physicists who considered Planck's original theory a too radical break with classical physics. For example, in an address of December 1912 Robert Millikan (1913, p. 123) judged the new theory to be "the most fundamental and the least revolutionary form of quantum theory, since it modifies classical theory only in the assumption of discontinuities in *time*, but not in *space*, in the emission (not in the absorption) of radiant energy."

Although the radiation was emitted with discrete energy values in Planck's theory, and all of the energy emitted at once, the oscillators did not possess intrinsically

---

[1] Most of Planck's articles on radiation and quantum theory are conveniently collected in Planck (1958). For historical studies of Planck's second theory, see Kuhn (1978, pp. 235–254), Needell (1980), and Mehra and Rechenberg (1982–2000, vol. 1, pp. 124–127, 146–150).

discontinuous energies. They could take on any energy, but the emission would only occur when the energy had reached values of $nh\nu$. In a letter to Paul Ehrenfest of 23 May 1913, Planck admitted his dislike of the discrete features of quantum theory: "I fear that your hatred of the zero-point energy extends to the electrodynamic emission hypothesis that I introduced and that leads to it. But what's to be done? For my part, *I hate discontinuity of energy even more than discontinuity of emission*" (Kuhn 1978, p. 253; emphasis added). As stated in the letter, a new and mysterious "zero-point energy" was part and parcel of Planck's new theory.

Based on the ideas underlying the second theory, Planck calculated the average energy of an oscillator vibrating with frequency ν to vary with the absolute temperature T as

$$\bar{E} = \frac{h\nu}{2} + \frac{h\nu}{\exp(h\nu/kT) - 1}, \tag{3.2}$$

where k is Boltzmann's constant. The values of the quantized energy levels of an oscillator could be written

$$\bar{E} = \tfrac{1}{2}(E_n - E_{n-1}) = \left(n + \tfrac{1}{2}\right)h\nu, \tag{3.3}$$

where $n = 0, 1, 2,...$ As Planck pointed out, these results imply that at $T = 0$ (or for $T \to 0$) the average energy is not zero but equals the finite energy $\frac{1}{2}h\nu$: "This rest-energy remains with the oscillator, on the average, at the absolute zero of temperature. It [the oscillator] cannot lose it, for it does not emit energy so long that $\bar{U}[=\bar{E}]$ is smaller than $h\nu$" (Planck 1911, p. 145). In order to derive the experimentally confirmed radiation law relating the energy density ρ of blackbody radiation to frequency and temperature, Planck appealed to the classical limit given by the Rayleigh-Jeans expression

$$\rho(\nu, T) = 8\pi \left(\frac{\nu}{c}\right)^3 \frac{kT}{\nu}, \tag{3.4}$$

where c is the speed of light in vacuum. By making use of a correspondence argument, he obtained in this way the same expression he had derived in 1900:

$$\rho(\nu, T) = \frac{8\pi h\nu^3/c^3}{\exp(h\nu/kT) - 1}. \tag{3.5}$$

According to Max Jammer, Planck's reasoning in 1911 "was probably the earliest instance in quantum theory of applying what more than 10 years later became known as the 'correspondence principle'" (Jammer 1966, p. 50; Whitaker 1985). Bohr formulated his correspondence principle in 1918, although in a limited sense it already appeared in his atomic theory of 1913. He introduced the name "correspondence principle" in 1920.

In his *Theorie der Wärmestrahlung*, Planck emphasized that the existence of a zero-point energy was completely foreign to classical physics. It seemed to be a ghost-like entity, on which it was difficult to get an experimental handle. As he noted in his first paper of 1911, since the new energy expression of an oscillator differed from the old one by only an additive constant, it would have no effect on the spectrum or on the specific heat as given by $c = \partial \bar{E}/\partial T$. For this reason, Walther Nernst's confirmation of Einstein's 1907 theory of the specific heat of solids could not be used to differentiate between the two radiation hypotheses. "Thus, so far it appears not really possible to make a direct experimental test of the new expression for $\bar{U}[=\bar{E}]$," he commented (Planck 1911, p. 146).

Planck similarly pointed out that Einstein's controversial theory of light quanta, or rather the photoelectric law derived from it, was unable to distinguish between the two hypotheses (Planck 1958, p. 284). Although he did not think of the zero-point energy as a measurable quantity, or one which would otherwise have direct experimental consequences, he did mention various phenomena that might justify it empirically in a qualitative sense. Among these phenomena was the experimental fact that the energy released in radioactive decay remained uninfluenced by even the most extreme cold. Moreover, the relativistic mass-energy equivalence $E = mc^2$ led naturally to the assumption of "a very considerable intra-atomic amount of energy also at zero absolute temperature" (Planck 1913, p. 140).

Planck's second quantum theory was short-lived, a major reason being its failure to comply with Bohr's atomic theory of 1913 in which both emission and absorption of radiation occurred discontinuously. The successful application of Bohr's theory to atomic and molecular spectroscopy argued against Planck's second theory, which nonetheless may have inspired Bohr in the development of his ideas of atomic structure (Kuhn 1978, p. 320). At any rate, Bohr soon came to the conclusion that Planck's notion of atomic oscillators was incompatible with his atomic theory. He eventually expressed misgivings about the Planckian oscillators because they were "inconsistent with Rutherford's theory, according to which all the forces between the particles of an atomic system vary inversely as the square of the distance apart" (Bohr 1913, p. 874). After all, as he said in a lecture in Copenhagen at the end of 1913, "No one has ever seen a Planck's resonator, nor indeed even measured its frequency of oscillation; we can observe only the period of oscillation of the radiation which is permitted."[2]

In an important but unpublished paper of 1916 Bohr emphasized that Planck's second theory was inconsistent with the basic assumption that an atomic system can exist only in a series of discrete stationary states. He argued that the probability of a quantum system being in a state n was given by

$$P_n = n^{r-1}, \tag{3.6}$$

where r denotes the number of degrees of freedom. For a system of several degrees of freedom ($r > 1$), the probability of the system being in state $n = 0$ should thus be

---

[2] Bohr (1922, p. 10), a translation of an address given to the Danish Physical Society on 20 December 1913 and published in Danish in *Fysisk Tidsskrift* 12 (1914), pp. 97–114.

zero. "Such a consideration gives a simple explanation of the mysterious zero-point energy," he wrote to the Swedish physicist Carl Wilhelm Oseen in a letter of 20 December 1915 (Bohr 1981, p. 567). In his unpublished paper he regained the result that at $T = 0$ a harmonic oscillator of two degrees of freedom would have a non-zero energy, but "This so-called zero-point energy has here an origin quite distinct from that in Planck's theory." Bohr elaborated: "In the present theory it arises simply from the fact, that... there is no probability of a periodic system of several degrees of freedom being in the state corresponding to $n = 0$.... At $T = 0$ all the systems are therefore in a state corresponding to $n = 1$" (Bohr 1981, p. 456; Gearhart 2010).[3]

Bohr not only applied his theory to the specific heat of hydrogen at low temperatures, but also to the quantized hydrogen atom he had introduced in his theory of 1913. According to this theory, the energy levels of the hydrogen atom were given by

$$E_n = -\frac{2\pi^2 e^4 m}{h^2} \frac{1}{n^2}, \tag{3.7}$$

where e is the charge of the electron and m its mass. Since the system has three degrees of freedom, $P_n = n^2$. Bohr explained: "This system affords a peculiar case of zero-point energy. Strictly there is no sense in considering the state corresponding to $n = 0$, since this would correspond to an infinite negative value for the energy; and in order to obtain agreement with experiments it must be assumed that the normal state of the system corresponds to $n = 1$" (Bohr 1981, pp. 459–460).

By about 1920 few physicists considered Planck's second theory a viable alternative. In a contribution to a special issue of *Die Naturwissenschaften* celebrating the 10-year anniversary of Bohr's atomic theory, Planck admitted that "This second formulation of the quantum theory may be considered today, at least in its extreme form, as finally disproved" (Planck 1923, p. 537). What persuaded him was the Stern-Gerlach experiment carried out by Otto Stern and Walther Gerlach in the early 1920s. The main outcome of this famous experiment was a proof of space quantization or the discreteness of the magnetic moment of atomic particles, but Planck and most other physicists also saw it as proof of the discrete stationary states postulated by Bohr's theory.

Yet, although the second theory had been abandoned by 1923, one element associated with it continued to live on: the zero-point energy. Planck had himself replaced his second quantum theory with a modified "third theory" (Needell 1980, pp. 249–268) and in this version the zero-point energy survived. In a letter of 10 March 1915 to the Dutch pioneer of low-temperature physics, Heike Kamerlingh Onnes in Leiden, he wrote: "I have almost completed an improved formulation of the quantum

---

[3] The unpublished paper, intended to appear in the April 1916 issue of *Philosophical Magazine*, was entitled "On the Application of the Quantum Theory to Periodic Systems." Due to Arnold Sommerfeld's new formulation of the quantum theory of atoms, Bohr decided to withdraw it shortly before it was to be published. Incidentally, Sommerfeld ignored the zero-point energy, which is not mentioned in any of the editions (1919–1924) of his influential book *Atombau und Spektrallinien* (Atomic Structure and Spectral Lines).

hypothesis applied to thermal radiation. I am more convinced than ever that zero-point energy is an indispensable element. Indeed, I believe I have the strongest evidence for it" (van Delft 2007, p. 491).

References

Bohr, N.: Niels Bohr. Collected Works. In: Hoyer, U., (ed.) vol. 2. Amsterdam, North-Holland (1981)
Bohr, N.: On the constitution of atoms and molecules. Philos. Mag. **26**, 1–25, 476–502, 851–875 (1913)
Bohr, N.: The Theory of Spectra and Atomic Constitution. Cambridge University Press, Cambridge (1922)
van Delft, D.: Freezing Physics: Heike Kamerlingh Onnes and the Quest for Cold. Koninklijke Nederlandse Akademie van Wetenschappen, Amsterdam (2007)
Gearhart, C.A.: "Astonishing successes" and "bitter disappointment": The specific heat of hydrogen in quantum theory. Arch. Hist. Exact Sci. **64**, 113–202 (2010)
Jammer, M.: The Conceptual Development of Quantum Mechanics. McGraw-Hill, New York (1966)
Kuhn, T.S.: Black-Body Theory and the Quantum Discontinuity, 1894–1912. Clarendon Press, Oxford (1978)
Mehra, J., H. Rechenberg.: The Historical Development of Quantum Theory, vol. 6. Springer, New York (1982–2000)
Millikan, R.A.: Atomic theories of radiation. Science **37**, 119–133 (1913)
Needell, A.A.: Irreversibility and the Failure of Classical Dynamics: Max Planck's Work on the Quantum Theory 1900–1915, p. 3058. University Microfilms, Ann Arbor, Michigan (1980)
Planck, M.: Eine neue Strahlungshypothese. Verh. Dtsch. Phys. Ges. **13**, 138–148. (Reprinted in Planck 1958, pp. 249–259) (1911)
Planck, M.: Über die Begründung das Gesetzes des schwarzen Strahlung. Ann. Phys. **37**, 642–656. (Reprinted in Planck 1958, pp. 287–301) (1912)
Planck, M.: Vorlesungen über die Theorie der Wärmestrahlung. J. A. Barth, Leipzig (1913)
Planck, M.: Die Bohrsche Atomtheorie. Die Naturwissenschaften **11**, 535–537 (1923)
Planck, M.: Physikalische Abhandlungen und Vorträge, vol. 2. Vieweg und Sohn, Braunschweig (1958)
Whitaker, M.A.B.: Planck's first and second theories and the correspondence principle. Eur. J. Phys. **6**, 266–270 (1985)

Chapter 4
Half-Quanta and Zero-Point Energy

Abstract During the first half of the 1920s the hypothesis of zero-point energy and the associated hypothesis of half-integral quantum numbers remained controversial. It was not until 1924 that they received solid confirmation from molecular spectroscopy. Although they could not be ignored any longer, they could not be understood on the basis of existing semi-classical quantum theory. Only with the emergence of quantum mechanics in 1925–1926 was the zero-point energy of material systems justified by a fundamental physical theory. The effect can be seen as a consequence of the uncertainty principle that Werner Heisenberg formulated in 1927.

Keywords Quantum theory · Zero-point energy · Molecules · Quantum numbers

The assumption of a zero-point energy attracted much attention in the physics community, although for more than a decade it remained uncertain whether the quantity was physically real or not (Milloni and Shih 1991; Mehra and Rechenberg 1999; van Delft 2007, pp. 484–493). Einstein was perhaps the first to come up with a physical argument for its existence, which he did in a paper of early 1913 co-authored by Otto Stern, a young physical chemist who had recently obtained his doctorate in Breslau under Otto Sackur and subsequently joined Einstein as his assistant, first in Prague and then in Zurich. Thirty years later he would receive the Nobel Prize in physics for his development of molecular and atomic beam methods of which the Stern–Gerlach effect was an early example.

In their work of 1913 Einstein and Stern considered the rotational energy of a diatomic molecule, as given by

$$E_{\mathrm{rot}} = \tfrac{1}{2} J (2\pi \nu)^2, \tag{4.1}$$

where J is the moment of inertia and ν the frequency of rotation. For a collection of molecules at fixed temperature they assumed that all molecules would rotate with the same speed. Moreover, they took the rotational kinetic energy to be twice as

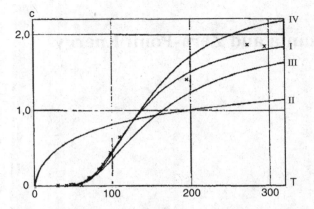

Fig. 4.1 The variation of the specific heat of hydrogen with the temperature according to Einstein and Stern (1913). The crosses are the experimental data. The theoretical curve I assumes a zero-point energy of $h\nu/2$ and curve IV of $h\nu$, whereas curve II assumes no zero-point energy

great as the kinetic energy of a one-dimensional oscillator vibrating at the same frequency ν, or equal to its average energy. From these assumptions they obtained different expressions for the specific heat $c = \partial E_{\text{rot}}/\partial T$, corresponding to both the first and the second of Planck's hypotheses. They then compared these theoretical results with the latest experimental measurements. Working at Nernst's laboratory in Berlin, Eucken (1912) had recently obtained data for molecular hydrogen at low temperatures that defied explanation in terms of existing theory. According to the calculations of Einstein and Stern, a fair agreement with Eucken's curve could be obtained if the zero-point energy $\frac{1}{2}h\nu$ were included, while Planck's first theory led to quite wrong results (Fig. 4.1). From this followed their cautious conclusion: "Eucken's results on the specific heat of hydrogen make probable the existence of a zero-point energy equal to $h\nu/2$" (Einstein and Stern 1913, p. 560).[1]

In spite of the appealing agreement between theory and experiment provided by the assumption of a zero-point energy, Einstein soon retracted his support of it. For one thing, Planck's second quantum theory presupposed harmonic oscillators at fixed frequencies, and there was no reason to expect that it would be applicable also to molecules rotating at frequencies depending on the temperature. Even more problematic was that Einstein and Stern were able to derive Planck's radiation law "without recourse to any kind of discontinuities"— assuming a zero-point contribution to the oscillator energy of $h\nu$ and not $\frac{1}{2}h\nu$! The confusion only increased when Paul Ehrenfest showed in a paper of 1913 that he could reproduce the data for low temperatures on the basis of statistical mechanics and Planck's first theory—that is, without any zero-point energy at all.[2]

---

[1] For context and analysis of the Einstein-Stern paper, see Milloni and Shih (1991), Mehra and Rechenberg (1999), Gearhart (2010) and Einstein (1995, pp. 270–273).

[2] Ehrenfest (1913). For historical analysis, see Klein (1970, pp. 264–273), Navarro and Perez (2006, pp. 215–223) and Gearhart (2010, pp. 135–138). In his unpublished paper of 1916, Bohr criticized

According to Ehrenfest, the quantum discontinuity was indispensable, whereas the zero-point energy was not. It was in this context that he assumed the angular momentum of the rotator to be quantized according to

$$\frac{1}{2} J (2\pi\nu)^2 = n \frac{h\nu}{2},$$ (4.2)

However, Ehrenfest did not formulate the quantization of angular momentum $L = nh/2\pi$ as a general principle, as Bohr would do independently a few months later.

Already in the fall of 1913 Einstein withdrew his support of the zero-point energy and the results reported in his paper with Stern. During the second Solvay conference in late October 1913 the question of the zero-point energy was discussed by Einstein, Wilhelm Wien, Nernst, and Lorentz. Einstein commented: "I no longer consider the arguments for the existence of zero-point energy that I and Mr. Stern put forward to be correct. Further pursuit of the arguments that we used in the derivation of Planck's radiation law showed that this road, based on the hypothesis of zero-point energy, leads to contradictions" (Einstein 1995, p. 553). In a letter to Ehrenfest a few days later he declared the zero-point energy "dead as a doornail" (*Mausetot*).[3] However, this announcement of death was premature.

Two points should be emphasized with regard to the paper by Einstein and Stern. First, they did not quantize the rotator, but allowed it to have a continuum of energies depending on the temperature. Second, they only attributed a zero-point energy to material objects, either oscillating electrons or rotating diatomic molecules, while they did not apply an additional energy term to the electromagnetic field. In retrospect, this explains how they were able to derive the correct Planck spectrum on the basis of the wrong zero-point energy $h\nu$. This value happens to be the correct one for the sum of the interacting harmonic oscillators and the energy of the electromagnetic field. The reasons that Einstein and Stern might have had for ignoring the zero-point energy of the field are discussed by Milloni and Shih (1991). Other shortcomings of the Einstein-Stern paper from a modern perspective are also pointed out by Dahl (1998) and Sciama (1991).

Far from being dead as a doornail, after 1913 the zero-point energy continued to attract a great deal of attention among physicists and physical chemists, in many cases independent of Planck's second theory. One phenomenon where it might be expected to manifest itself was the X-ray diffraction pattern in crystals at low temperature. If the atoms in a crystal had a zero-point motion, this would presumably influence the intensity distribution in the diffraction pattern. This line of research, which eventually led to a "direct proof" of zero-point motion, was pioneered by Peter Debye during

---

Ehrenfest's theory and found a better fit to the data on $c(T)$ from his own theory (Bohr 1981, pp. 458–460).

[3] Einstein to Ehrenfest, before 7 November 1913, and also Einstein to Ludwig Hopf, 2 November 1913: "One hopes Debije [Debye] will soon demonstrate the incorrectness of the hypothesis of zero-point energy, the theoretical untenability of which became glaringly obvious to me soon after the publication of the paper I coauthored with Mr. Stern." Both letters are reproduced in Einstein (1993, pp. 563–565).

1913–1914, but at the time without leading to a conclusive answer (Debye 1913, 1914). The proof was only obtained in the late 1920s, based on quantum-mechanical analysis of diffraction patterns from NaCl crystals at low temperatures. Reginald James et al. (1928) concluded that their calculations "agree very closely with the experimental curves on the assumption that the crystal possesses zero-point energy of amount half a quantum per degree of freedom, as proposed by Planck."

Another line of research was related to the attempts to separate isotopes by chemical means or fractional distillation. The British physicist Fredrick Lindemann, a former collaborator of Nernst, showed that in principle such separation would be possible, but that it would depend on whether or not there was a zero-point energy: "The amount of separation to be expected depends upon ... whether 'Nullpunktsenergie' is assumed. ... The difference should be measurable if there is no 'Nullpunktsenergie', and it is suggested that experiments on the vapour pressure and affinity of isotopes would give valuable information on this important point" (Lindemann 1919, p. 181). If a zero-point energy were not assumed, the expected separation effect would be tiny. However, experiments of the kind proposed were unable to settle the question and tell whether the zero-point energy existed or not.

Arguments somewhat similar to Lindemann's were suggested a few years later by Stern, who discussed them with a sceptical Wolfgang Pauli. In a letter to Charles Enz (1974, 2002, p. 150) of 21 January 1960, Stern recalled:

> Pauli and I continually discussed the question of the zero-point energy in Hamburg in the early 1920s. ... I for my part always tried to convert Pauli to the zero-point energy against which he had the gravest hesitations. My main argument was that I had calculated the vapour pressure differences of the neon isotopes 20 and 22, which Aston had tried in vain to separate by distillation. If one calculates without zero-point energy there results such a large difference that the separation should have been quite easy. The argument seemed (and seems) to me so strong because one does not assume anything else than Planck's formula and the fact that isotopes are distinguished only by the atomic weight.

Stern did not publish his calculations on the isotopic effect. In any case, at the time Pauli remained unconvinced. As mentioned by Stern, as early as 1913 he had studied the vapor pressure of monatomic gases and arrived at an expression for the heat of vaporization which he interpreted as support of a zero-point energy (Stern 1913). He gave a more elaborate version in a work of 1919, in which he calculated the vapor pressure above the surface of a solid body, which he conceived as a collection of N atoms vibrating harmonically in three dimensions with frequencies ν_k. To obtain agreement with experimental data he suggested that the heat of vaporization at $T = 0$ was smaller than the potential energy of the N atoms in the gaseous state. That is, in the solid equilibrium state the atoms were not at rest, but possessed a vibrational energy of

$$E_0 = \tfrac{1}{2} h \sum_{k=1}^{3N} \nu_k. \qquad (4.3)$$

Stern (1919, p. 77) hoped that the zero-point energy of solid bodies would find its interpretation in "The more recent works of N. Bohr [in which] this hypothesis in a somewhat modified form has acquired a very deep meaning." It is unclear which works of Bohr Stern had in mind, but he most likely thought of the new correspondence principle.

Rather than considering vaporization, the two young physical chemists Kurt Bennewitz and Franz Simon (later Sir Francis Simon), who worked at Nernst's laboratory in Berlin, studied the melting process at low temperatures. Their complex calculations of the melting points of hydrogen, argon and mercury led them to conclude that the results provided evidence for a zero-point energy. Moreover, they suggested—correctly, as it later turned out—that this quantity was responsible for the difficulty in solidifying helium even at very low temperature. According to Bennewitz and Simon (1923) the zero-point energy in liquid helium would act as an internal pressure, expanding it to such a low density that no rigid structure of the atoms could be maintained.

Among the early and most persistent advocates of the zero-point energy was the Dutch physicist Willem Keesom, who at the 1913 Wolfkehl meeting in Göttingen defended the new Einstein-Stern theory and suggested that the zero-point energy might also turn up in the equation of state of monatomic gases (Planck et al. 1914, pp. 166, 194; Keesom 1913). Several other speakers at the Wolfkehl meeting commented on the zero-point energy, including Planck, Kamerlingh Onnes, Debye, and Sommerfeld. For a while the subject was taken seriously among Dutch physicists in Leiden and Utrecht, where it came up in particular in connection with research in magnetism at low temperature (van Delft 2007, pp. 484–493; 2008). As Keesom saw it, the evidence in favor of zero-point energy was far stronger than the counterevidence. Yet, evidence is not proof, and in the decade after 1911 the problem remained unresolved. As mentioned, the abandonment of Planck's second theory did not imply that the idea of zero-point energy was also abandoned.

Einstein would have none of it. "It is well known that all theories characterized by a 'zero-point energy' face great difficulties when it comes to an exact treatment," he wrote in a paper of 1915. "No theoretician," he continued, "can at present utter the word 'zero-point energy' without breaking into a half-embarrassed, half-ironic smile" (Einstein 1915, p. 237). Yet several years later Einstein returned to the question, now with a more sympathetic view. In his correspondence with Ehrenfest from 1921–1923 he suggested that the zero-point energy might play a role in the cases of hydrogen and helium. Perhaps, he suggested, it might explain the maximum density in helium (Einstein 2009, p. 265; Mehra and Rechenberg 1982–2000, vol. 1, pp. 571–572). However, neither Einstein nor Ehrenfest turned their ideas on the subject into publications.

According to Bohr's atomic theory quantum numbers had to be integers, but by the early 1920s a growing amount of evidence indicated that in some cases "half-quanta" of the kind first considered by Planck in 1911 had to be accepted. These half-integral quantum numbers first turned up in attempts to understand the band spectra emitted by molecules. In 1919 Elmer Imes at the University of Michigan published precision experiments on the absorption of HCl and HBr that showed a distinct gap in the centre of the pattern of lines (Imes 1919; Gearhart 2010). In order to explain Imes's data, the Berlin physicist Fritz Reiche suggested changing the standard rule for rotational quantization, namely by changing the formula for the energy of a rotator from

$$E_{\text{rot}} = m^2 \frac{h^2}{8\pi^2 J} \qquad \text{to} \qquad E_{\text{rot}} = \left(m + \tfrac{1}{2}\right)^2 \frac{h^2}{8\pi^2 J}. \tag{4.4}$$

Since the quantum number $m = 0, 1, 2, \ldots$, the change implied a zero-point rotational energy. This conclusion, that a diatomic molecule cannot exist in a rotation-free state, he justified by Bohr's new correspondence principle. The same result, also based on the correspondence principle, was derived by Kramers and Pauli (1923). According to Reiche (1920, 1921, pp. 155–159), the suggestion of rotational half-quanta was first made by Einstein, "with whom I have often had the opportunity to discuss these matters, ... [and who mentioned] a possible way to change the rotational quantization so as to annul the contradiction with observations." Although the half-quanta were theoretically controversial they seemed necessary and were adopted by several molecular spectroscopists. For example, they were incorporated into an influential and more elaborate theory of band spectra that Adolf Kratzer, a physicist at the University of Münster, published in 1923 (Kratzer 1923; Barker 1923).

With the new studies of band spectra in the early 1920s the concept of zero-point energy became respectable among molecular physicists. Yet it was only in the fall of 1924 that half-quanta were firmly established in molecular spectroscopy. In a study of the spectrum of boron monoxide (BO), Robert Mulliken, a young physical chemist at Harvard University, concluded that observations could only be understood on the assumption of quantum numbers with a minimum value of one-half. In a preliminary announcement of his results in *Nature*, he wrote (Mulliken 1924):

> It is probable that the minimum vibrational energy of BO (and doubtless of other) molecules is $\frac{1}{2}$ quantum. In the case of molecular rotational energy, the necessity of using half quanta is already well established. Analogous relations appear in line spectra; e.g. Heisenberg has successfully used half-integral radial and azimuthal quantum numbers in explaining the structure and Zeeman effect of doublets and triplets.

In the full report that appeared in *Physical Review* in March 1925, Mulliken (1925, p. 281) similarly concluded that his work "would involve a null-point energy of $\frac{1}{2}$ quantum each of vibration and rotation," and he related it to Lindemann's investigation of the vapor pressures of isotopes. His paper was widely considered a final confirmation of half-quanta and, by implication, a form of zero-point energy. On the other hand, in spite of being anomalous the result had almost no effect at all on the crisis in quantum theory that would lead a few months later to Heisenberg's formulation of a new quantum mechanics—and thereby to a theoretical justification of the zero-point energy of an oscillator.

At this point a brief terminological note may be appropriate. What Planck had originally called *Restenergie* (rest energy) soon became known as *Nullpunktsenergie*, a name used by, for example, Einstein and Stern in their paper of 1913. For a while the German term—or sometimes the equivalent "null-point energy" as used by Mulliken—appeared also in the English scientific literature, as for instance in paper by Lindemann (1919) and Tolman (1920). Only from about 1925 did it become common to refer to "zero-point energy." This term may first have been used by Bohr in his unpublished paper of 1916.

As indicated in the quotation from Mulliken, half-quanta also played a role in some of the attempts to understand what was probably the most serious problem in the old quantum theory, namely, the anomalous Zeeman effect. Thus, according to the young Heisenberg's so-called core model of the atom, electrons could be in a state given by the azimuthal quantum number $k = \frac{1}{2}$, which was difficult to reconcile with the established Bohr-Sommerfeld atomic model (Cassidy 1978). When Heisenberg introduced half-integral quantum numbers, he was originally unaware of the earlier discussion related to Planck's second theory and the possibility of a zero-point energy. It seems to have been Pauli who directed his attention to this discussion and to Stern's paper of 1919, and Heisenberg also had conversations with Kratzer who informed him about the use of half-quanta in the study of band spectra (Mehra and Rechenberg 1982–2000, vol. 2, p. 30; Gearhart 2010). To make a long story short, in spite of resistance from Bohr and other leading physicists the evidence for half-quanta and zero-point energy could not be ignored: physicists learned to live with them, if not love them.

It was only with the emergence of quantum mechanics that the concept of zero-point energy became really respectable and seen as a consequence of a fundamental physical theory. In his famous *Umdeutung* paper, which marks the beginning of modern quantum mechanics, Heisenberg (1925) applied his new formalism to the harmonic oscillator, the result being

$$E_n = \left(n + \tfrac{1}{2}\right) h\nu . \tag{4.5}$$

For the anharmonic oscillator he derived a more complicated expression, also involving a zero-point energy. Although Heisenberg's result was the same as Planck's formula from 1911, there was the difference that in the case of quantum mechanics it is valid also for an individual oscillator and not merely as an average. This result was duplicated by Erwin Schrödinger in the second of his communications on wave mechanics from April 1926, where he commented: "Strangely, our quantum levels are precisely the same as in Heisenberg's theory!" (Schrödinger 1926, p. 516). The formal equivalence between wave mechanics and the Göttingen quantum mechanics was only proved a month later. Unlike Heisenberg, Schrödinger noted the connection to the old question of the validity of Planck's second quantum theory.

By the summer of 1926 the zero-point energy was no longer controversial, at least not insofar as it concerned material systems. In the later literature on quantum physics it became customary to see the zero-point energy as a straightforward consequence of Heisenberg's uncertainty principle for position and momentum, in one dimension and with $\hbar \equiv h/2\pi$,

$$\Delta p_x \Delta x \geqslant \hbar. \qquad (4.6)$$

If a harmonic oscillator were to have zero energy, both its potential and kinetic energy would have to be zero. The case $E_{\text{pot}} = \frac{1}{2}kx^2 = 0$ would correspond to a precise knowledge of the position of the particle ($x = 0$) and thus imply that the momentum p is completely uncertain. However, then the mean value of $E_{\text{kin}} = p^2/2m$ would be infinite. Conversely, for E_{kin} to be zero, E_{pot} would have to be infinite. A simple calculation shows that the ground state of a quantum harmonic oscillator is equal to the minimum energy allowed by the uncertainty principle, and that this energy is just $\frac{1}{2}h\nu$.

References

Barker, E.F.: Molecular spectra and half-quanta. Astrophys. J. **58**, 201–207 (1923)

Bennewitz, K., Simon, F.: Zur frage der nullpunktsenergie. Zeitschrift für Physik **16**, 183–199 (1923)

Bohr, N.: Niels Bohr. Collected Works. In: Hoyer, U. (ed.) vol. 2. Amsterdam, North-Holland (1981)

Cassidy, D.C.: Heisenberg's first paper. Phys. Today **31**(July), 23–28 (1978)

Dahl, J.P.: On the Einstein-Stern model of rotational heat capacities. J. Chem. Phys. **109**, 10688–10691 (1998)

Debye, P.: Über den Einfluss der Wärmebewegung auf die Interferenzerscheinungen bei Röntgenstrahlungen. Verh. Dtsch. Phys. Ges. **15**, 678–689 (1913)

Debye, P.: Interferenz von Röntgenstrahlungen und Wärmebewegung. Ann. Phys. **43**, 49–95 (1914)

van Delft, D.: Freezing Physics: Heike Kamerlingh Onnes and the Quest for Cold. Koninklijke Nederlandse Akademie van Wetenschappen, Amsterdam (2007)

van Delft, D.: Zero-point energy: the case of the Leiden low-temperature laboratory of Heike Kamerlingh Onnes. Ann. Sci. **65**, 339–362 (2008)

Ehrenfest, P.: Bemerkung betreffs der spezifischen Wärme zweiatoniger Gase. Verh. Dtsch. Phys. Ges. **15**, 451–457 (1913)

Einstein, A.: The Collected Papers of Albert Einstein. Buchwald D.K. et al. (ed.) Vol. 12. Princeton: Princeton University Press (2009)

Einstein, A.: The Collected Papers of Albert Einstein. In: Klein, M.J., Kox, A.J., Schulmann, R., (ed.) vol. 5. Princeton: Princeton University Press (1993)

Einstein, A.: The Collected Papers of Albert Einstein. In: Klein M.J. et al. (ed.) Vol. 4. Princeton: Princeton University Press (1995)

Einstein, A., Stern, O.: Einige Argumente für die Annahme einer molekularen Agitation beim absoluten Nullpunkt. Ann. Phys. **40**, 551–560 (1913)

Einstein, A.: Experimenteller Nachweis der Ampèreschen Molekularströme. Die Naturwissenschaften **3**, 237–238 (1915)

Enz, C.P.: Is the zero-point energy real? In: Enz, C.P., Mehra, J. (eds.) Physical Reality and Mathematical Description, pp. 124–132. Reidel, Dordrecht (1974)

Enz, C.P.: No Time to Be Brief: A Scientific Biography of Wolfgang Pauli. Oxford University Press, Oxford (2002)

Eucken, A.: Die Molekularwärme des Wasserstoffs bei tiefen Temperaturen. Sitzungsberichte, Preussische Akademie der Wissenschaften, Berlin, pp. 141–151 (1912)

Gearhart, C.A.: Astonishing successes and bitter disappointment: the specific heat of hydrogen in quantum theory. Arch. Hist. Exact Sci. **64**, 113–202 (2010)

Heisenberg, W.: Über quantentheoretische Umdeutung kinematischer und mechanischer Beziehungen. Zeitschrift für Physik **33**, 879–893 (1925)

Imes, E.S.: Measurement in the near infrared absorption of some diatomic gases. Astrophys. J. **50**, 251–276 (1919)

James, R.W., Waller, I., Hartree, D.R.: An investigation into the existence of zero-point energy in the rock-salt lattice by an X-ray diffraction method. Proc. R. Soc. A **118**, 334–350 (1928)

Keesom, W.H.: On the equation of state of an ideal monatomic gas according to the quantum-theory. Communications from the Physical Laboratory of the University of Leiden 30b (1913)

Klein, M.J.: Paul Ehrenfest, vol. 1. The Making of a Theoretical Physicist, Amsterdam, North-Holland (1970)

Kramers, H.A., Pauli, W.: Zur Theorie der Bandenspektren. Zeitschrift für Physik **13**, 351–367 (1923)

Kratzer, A.: Die Feinstruktur einer Klasse von Bandenspektren. Ann. Phys. **71**, 72–103 (1923)

Lindemann, F.: Note on the vapour pressure and affinity of isotopes. Phil. Mag. **38**, 173–181 (1919)

Mehra, J., Rechenberg, H.: The Historical Development of Quantum Theory, Vol. 6. Springer, New York (1982–2000)

Mehra, J., Rechenberg, H.: Planck's half-quanta: a history of the concept of zero-point energy. Found. Phys. **29**, 91–132 (1999)

Milloni, P.W., Shih, M.-L.: Zero-point energy in early quantum theory. Am. J. Phys. **59**(August), 684–697 (1991)

Mulliken, R.S.: The band spectrum of boron monoxide. Nature **114**, 349–350 (1924)

Mulliken, R.S.: The isotope effect in band spectra, II: the spectrum of boron monoxide. Phys. Rev. **25**, 259–294 (1925)

Navarro, L., Pérez, E.: Paul Ehrenfest: the genesis of the adiabatic hypothesis, 1911–1914. Arch. Hist. Exact Sci. **60**, 209–267 (2006)

Planck, M., et al.: Vorträge über die Kinetische Theorie der Materie und der Elektrizität. B. G. Teubner, Leipzig (1914)

Reiche, F.: Zur Theorie der Rotationsspektren. Zeitschrift für Physik **1**, 283–293 (1920)

Reiche, F.: Die Quantentheorie: Ihr Ursprung und ihre Entwicklung. Julius Springer, Berlin (1921)

Schrödinger, E.: Quantisierung als Eigenwertproblem. Zweite Mitt. Ann. der Physik **79**, 489–527 (1926)

Sciama, D.W.: The physical significance of the vacuum state of a quantum field. In: Saunders, S., Brown, H.R. (eds.) The Philosophy of Vacuum, pp. 137–158. Clarendon Press, Oxford (1991)

Stern, O.: Zur kinetischen Theorie des Dampfdrucks einatomiger freier Stoffe und über die Entropiekonstante einatomiger Gase. Phys. Z. **14**, 629–632 (1913)

Stern, O.: Zusammenfassender Bericht über die Molekulartheorie des Dampfdruckes fester Stoffe und ihre Bedeutung für die Berechnung chemischer Konstanten. Zeitschrift für Electrochemie **25**, 66–80 (1919)

Tolman, R.C.: The entropy of gases. J. Am. Chem. Soc. **42**, 1185–1193 (1920)

Chapter 5
Nernst's Cosmic Quantum-Ether

Abstract According to a theory proposed by Walther Nernst in 1916, empty space (or the ether) was a reservoir of zero-point electromagnetic radiation with an energy density of the order 10^{23} erg cm^{-3}. Nernst's idea implied a new picture of empty space in which heavy atomic nuclei might emerge from vacuum fluctuations, and he would later develop it into a speculative cosmological theory. Although a few physicists found it interesting, the majority either ignored it or dismissed it as unfounded. Only much later has this early idea been reevaluated and seen as an anticipation of the vacuum of modern quantum physics.

Keywords Ether · Zero-point energy · Radiation energy density · Walther Nernst

The first suggestion of applying the concept of zero-point energy to free space, and in this way turning it into a tool of possible relevance for cosmological research, came from an unlikely source. The great physical chemist Walther Nernst had established his reputation by pioneering works in electrochemistry and chemical thermodynamics, culminating in 1906 with the heat theorem also known as the third law of thermodynamics (Fig. 5.1).[1] It was primarily for his work in thermodynamics that he was awarded the 1920 Nobel Prize in chemistry. The new heat theorem led Nernst from chemistry to quantum physics, a move inspired by Einstein's 1907 theory of the specific heats of solids, which Nernst confirmed in a series of low-temperature experiments conducted about 1910.

Nernst's debut in quantum theory took place in early 1911, when he submitted a paper on the theory of specific heats in which he applied quantum theory to diatomic gases such as hydrogen. Although Nernst did not quantize the rotating molecule, he did arrive at a quantum-based phenomenological expression for the variation of the specific heat of a diatomic gas with temperature (Nernst 1911; Gearhart 2010). Later

---

[1] The literature on Nernst is extensive. For full biographies, see in particular Barkan (1999) and Bartel and Huebener (2007). The scientific works of Nernst are well discussed in Partington (1953). A useful website, with a bibliography of the works of Nernst and his students, is http://www.nernst.de.

H. S. Kragh and J. M. Overduin, *The Weight of the Vacuum*, SpringerBriefs in Physics, DOI: 10.1007/978-3-642-55090-4_5, © The Author(s) 2014

Fig. 5.1 Etching of Walther Nernst (1864–1941) at the time he developed his ideas of a universe filled with zero-point radiation

the same year he reported on his formula and related subjects at the memorable first Solvay conference, a meeting of which he was the chief organizer. In Brussels he listened to Planck's exposition of his second quantum hypothesis and its associated concept of zero-point energy. As Nernst suggested in the subsequent discussion, the zero-point energy would imply that at the absolute zero of temperature a solid body would still have a vapor pressure, a claim that Planck however denied (Langevin and de Broglie 1912, p. 129).

At about the same time that Nernst turned his attention to quantum theory, he took up an interest in cosmological questions (Browne 1995; Kragh 1995; Bartel and Huebener 2007, pp. 306–326). His first excursion into cosmic physics was not motivated by quantum theory, but by the old and still much-discussed question of a universal *Wärmetod* (heat death) caused by the ever increasing amount of entropy. This supposed consequence of the second law of thermodynamics, first stated in different versions by Hermann von Helmholtz, William Thomson and Rudolf Clausius in the mid-nineteenth century, was highly controversial for both scientific and non-scientific reasons because it predicted the end of the world, or at least the end of all activity and life in the world (Kragh 2008). Nernst's Swedish colleague in physical chemistry, Svante Arrhenius, was among those who in the early years of the twentieth century resisted the heat death scenario and suggested cosmic mechanisms that would counter the deadly growth in entropy. Probably inspired by Arrhenius's writings, Nernst did the same. Ever since the 1880s, when he first became acquainted with the gloomy predictions of an unavoidable cosmic heat death, he sought to refute this alleged consequence of the second law of thermodynamics. For him, as for several of his colleagues in science, it was an intellectual necessity to establish a cosmology that secured eternal evolution in an infinite, self-perpetuating universe.

In a lecture given to the 1912 meeting of the Society of German Scientists and Physicians in Münster, Nernst indicated a way in which the world might be saved from the heat death without abandoning the second law of thermodynamics. His tentative solution involved as major ingredients radioactivity and the ether, with the latter supposed to be the ultimate end product of radioactive decay. (As mentioned in Chap. 2, a similar idea had earlier been entertained by LeBon and other authors.) Like free energy, radioactivity was known to decrease irreversibly. Alone, it would not counter the entropy increase and prevent what Nernst dramatically called the *Götterdämmerung des Weltalls* (cosmic twilight of the Gods). On the contrary, "the theory of radioactive decay of the elements has augmented the above-mentioned degradation of energy with a correspondingly steady degradation of matter, and thus has only doubled the prospects of an Armaggeddon of the universe" (Nernst 1912, p. 105; Huber and Jaakkola 1995).

Nernst was not the first to use radioactivity in a cosmological context. A year earlier the Austrian physicist Arthur Erich Haas (1912) had reached the same conclusion, that radioactive decay constituted one more argument for the end of the universe. But whereas this was a conclusion Haas happily welcomed, Nernst sought to circumvent it and turn it into an argument for a static and eternally active world. This is where the ether entered, namely, as a medium that *ex hypothesi* counteracted the degradation of matter and energy. According to Nernst (1912, pp. 105–106):

> The atoms of all elements of the universe will sooner or later entirely dissolve in some primary substance [*Ursubstanz*], which we would have to identify with the hypothetical medium of the so-called luminiferous ether. In this medium ... all possible configurations can presumably occur, even the most improbable ones, and consequently, an atom of some element (most likely one with high atomic weight) would have to be recreated from time to time. ... This means, at any rate, that the cessation of all events no longer needs to follow unconditionally as a consequence of our present view of nature.

Nernst's cosmic hypothesis was admittedly speculative, and he emphasized that it should not be taken as a new cosmological theory but merely as an illustration of what he called "the thermodynamic approach" to the study of the universe. His chief hypothesis, an active ether in constant interaction with matter, was not particularly novel, and in 1912 he did not refer to either quantum theory or zero-point energy. However, in an article four years later he did make the connection. In this paper, a lengthy communication to the German Physical Society in 1916, Nernst proposed that empty space (or ether, as he saw it) was filled with electromagnetic zero-point radiation. Although he found the zero-point energy useful for the energy-rich ether, he was not satisfied with Planck's version of it because he felt it violated the universal validity of electrodynamics. His own alternative, he emphasized, "succeeds in taking over without changes the most important laws of the old physics, [and this] I consider not only an advantage but also a probable reason for admitting it as acceptable" (Nernst 1916, p. 107). Contrary to Planck and other early researchers, Nernst's zero-point energy characterized both material objects (oscillators and rotators) and the radiation filling the ether: "Even without the existence of radiating matter, that is, matter heated above absolute zero or somehow excited, empty space—or, as we prefer to say, the luminiferous ether—is filled with radiation" (p. 86). The two were interconnected, for a vibrating electron would constantly exchange energy with the zero-point radiation of the ether.

Another difference between the systems of Planck and Nernst was that energy conservation was strictly enforced in the former, but was valid according to Nernst only in a statistical sense, "just like the second law of thermodynamics." For a single atom or molecule the energy did not need to be conserved, since the material object would exchange energy with the hidden energy pool of empty space. This was a conception to which Nernst would return a few years later, extending it to the general suggestion that *all* the laws of nature were of a statistical nature. This was a radical idea, but not shocking to contemporary physicists facing the problems of understanding the quantum world. At about the same time Bohr and a few of his colleagues (including Hendrik A. Kramers and Charles G. Darwin, a grandson of the biologist) contemplated the possibility that strict energy conservation might break down in the interaction between matter and radiation.

In his inaugural lecture as rector for the University of Berlin, Nernst (1922) not only discussed the inherent uncertainty of the laws of nature, he also repeated his idea that an enormous amount of energy was stored in the light ether in the form of zero-point energy. Nernst's ether was quasi-material, in the sense that he imagined it to consist of tiny "molecules," which he conceived as neutral doublets made up of two polar particles of "unbelievably small dimensions." This idea of a corpuscular ether was not central to his arguments, however. He merely seems to have reused an older idea of his, namely that the ether consists of weightless combinations of positive and negative electrons. These hypothetical particles he called "neutrons."[2]

It was an important part of Nernst's hypothesis that calculations of the zero-point energy followed from the ordinary theory of statistical mechanics by means of the formal substitution

$$kT \to h\nu . \qquad (5.1)$$

[2] Nernst (1916, p. 110). Nernst's neutronic ether appeared as late as in the 15th edition of his widely used textbook in theoretical chemistry, where he stated that the electrons making up the neutrons would become ponderable by taking up zero-point energy (Nernst 1926, p. 464). He may have taken the idea, as well as the name "neutron," from the Australian physicist William Sutherland (1899). Nernst's neutron had only the name in common with the neutron that Ernest Rutherford introduced in 1920 as a material proton-electron composite particle.

This implies that for each degree of freedom, where classical theory assigns the energy $\frac{1}{2}kT$ the zero-point energy becomes $\frac{1}{2}h\nu$. For example, the ground state of a one-dimensional oscillator becomes $h\nu$ and not, as in Planck's theory, $\frac{1}{2}h\nu$ (Nernst 1916, p. 87; Peebles and Ratra 2003, p. 571). He commented: "Every atom, and likewise every conglomerate of atoms, which is capable of oscillation at a frequency ν per second owing to its mechanical conditions, will per degree of freedom take up the kinetic energy $E = \frac{1}{2}h\nu$ and that even, as already noted, at the absolute zero. ... Contrary to the usual heat motion, but in accordance with thermodynamics, the zero-point energy is, like every other form of energy at absolute zero, free energy" (Nernst 1916, pp. 86–87). For the energy density of zero-point radiation at frequency ν, Nernst adopted the formula

$$\rho(\nu, T) = \frac{8\pi h}{c^3} \nu^3 , \tag{5.2}$$

which derives from the classical Rayleigh-Jeans law by replacing kT with $h\nu$. The total energy density integrated over all frequencies then diverges (becomes infinite). Although Nernst saw "no reason to call such a conception impossible," of course he realized that an infinite energy density is unphysical. Based on his idea of an atomistic ether, he therefore considered replacing the ν^3 law with the expression

$$\rho(\nu, T) = \frac{8\pi h\nu^3}{c^3\nu_0} \frac{\nu}{\exp(\nu/\nu_0) - 1} , \tag{5.3}$$

where ν_0 is a constant characteristic of the structure of the ether-vacuum. However, given the lack of knowledge of the value of ν_0 he chose to return to the ν^3 law and provide it with a cut-off corresponding to some maximum frequency ν_m. The result becomes

$$\rho_{\text{vac}} = \int_0^{\nu_m} \frac{8\pi h}{c^3} \nu^3 d\nu = \frac{2\pi h}{c^3} \nu_m^4 . \tag{5.4}$$

As a numerical example, Nernst adopted a value $\nu_m = 10^{20}$ Hz, or $\lambda_{\text{min}} = 3 \times 10^{-10}$ cm, and with this obtained a lower limit for the energy density, namely

$$\rho_{\text{vac}} = 1.5 \times 10^{23} \text{ erg cm}^{-3} , \tag{5.5}$$

or, by $E = mc^2$, the equivalent of about 150 g cm^{-3}. "The amount of zero-point energy in the vacuum is thus quite enormous, causing ... fluctuations in it to exert great actions," he wrote (p. 89). Referring to a result obtained by Planck for the energy density of heat radiation, Nernst further showed that if the zero-point radiation enclosed in a container is compressed, neither its energy density nor its spectral distribution will be affected: "Any objections one might raise to the zero-point radiation owing to radiation pressure or resistance to bodies moving through the vacuum are overcome by this truly remarkable [gewiss merkwürdige] result" (p. 90). The remarkable result relied on the relationship $\rho \sim \nu^3$. The invariance of this energy density would later reappear as a property of the "false vacuum" of inflation cosmology and, even later, of dark energy.

Although it is Nernst's cosmophysical speculations based on an ethereal zero-point energy that are of interest in the present context, these ideas played only a limited role in his 1916 essay. The latter was mainly concerned with more mundane applications, in particular chemical reaction rates, equilibrium processes, and the structure of the hydrogen molecule. Based on his zero-point version of quantum theory he proposed a model of the hydrogen molecule that differed in some respects from the Bohr-Debye model generally accepted at the time (Bohr 1913; Debye 1915). According to this model, which was essentially the one proposed by Bohr (1913), the two revolving electrons were placed across from each other on a circular orbit perpendicular to and midway between the two hydrogen nuclei. Nernst ascribed to each of the two electrons a kinetic energy of $\frac{1}{2}h\nu$, where ν is the frequency of revolution. Because the electrons were in equilibrium with the zero-point radiation, they would not radiate, which explained the stability of the model without sacrificing the validity of ordinary electrodynamics as postulated by Bohr. For the moment of inertia of the hydrogen molecule Nernst derived $J = 3.6 \times 10^{-41}$ g cm^2, which he found was in better agreement with measurements than the value used by Debye (1.2×10^{-40} g cm^2).

In a booklet of 1921 entitled *Das Weltgebäude im Lichte der neueren Forschung* (The World Structure in the Light of Modern Research) and based on a popular lecture he gave in Berlin, Nernst elaborated on the cosmological and astrophysical consequences of his hypothesis. His larger aim was the same, to demonstrate that eternal matter-ether recycling prevented the heat death and secured a static universe without beginning or end: "Our eyes need not, in the far future, have to look at the world as a horrible graveyard, but as a continual abundance of brightly shining stars which come into existence and disappear" (Nernst 1921, p. 37; Bromberg 1976, pp. 169–171).

More clearly than before, Nernst explained that atoms of the chemical elements appeared out of the fluctuations of the ether, and that these atoms or their decay products would again disappear in the zero-point energy of the ethereal sea. This idea also appeared in several of his later works, where he attempted to develop it into a proper theory of astro- and cosmophysics. For example, in 1921 he considered the temperature of cosmic space, as usual identifying empty space with the ether. Without providing a value for the very low temperature, he argued that the ether must have a small capacity for absorbing heat rays and that the absorption of heat would eventually turn up as zero-point energy in the ether. He developed this theme in a series of later works. In 1938 he arrived at a cosmic "background temperature" of about 0.75 K, a result he considered to be "not implausible" (Nernst 1938). But we shall not be concerned here with Nernst's cosmological views in the 1930s or with his attempt to interpret Hubble's law of expansion as a quantum effect in a stationary universe (for discussion of these topics see Kragh 1995 or Bartel and Huebener 2007).

As Nernst pointed out in his *Weltgebäude* of 1921, the German physicist Emil Wiechert, a pioneer of geophysics and electron theory, had independently arrived at a view of the universe that was similar in many ways to his own. Wiechert was as committed to the ether as Nernst, and his ether was no less physically active and rich in energy. On the other hand, it was also a kind of vacuum, the medium left after the removal of all matter. Like Nernst, he speculated that ether-matter transmutations might continually take place in the depths of space, and in this way provide a cosmic cycle that would make the heat death avoidable. According to Wiechert, material atoms were to be seen as extraordinary configurations in the ether, which had to be assigned some energy content. "With regard to the structure of the electron," he wrote, "it follows that the energy density of the ether must be considered to be comparable to at least 7×10^{30} erg/cm^3. ... One gets an impression of the forces that govern the ether when one recalls that the pressure which comes into play by keeping together the electric charge in an electron is of the order 7×10^{24} atmospheres" (Wiechert 1921a, p. 66; 1921b). He was referring to the so-called Poincaré stress, which we shall meet in Chap. 7.

Whereas Wiechert did not follow Nernst in making use of the zero-point energy, or otherwise refer to quantum theory, he was aware of Nernst's ideas (Wiechert 1921b, p. 186). Moreover, he related the energy of the ether to the cosmological constant appearing in Einstein's field equations of 1917. (Nernst ignored general relativity and never mentioned the cosmological constant or Einstein's world model.) Although strongly opposed to the theory of relativity, in large measure because it disposed of his beloved ether, Wiechert (1921a p. 69) suggested that the general theory had in effect resurrected the ether and that the cosmological constant (Λ or λ) somehow played a role in the resurrection. In an interesting passage he wrote:

My impression is that the λ-term does not subordinate the ether to matter, but, on the contrary, subordinates matter to ether; for now matter appears as precipitations from the ether which here and there are rolled up and thereby cause insignificant changes in the constitution of the ether.

Unfortunately, he was not more concrete than that.

Nernst's speculations had some similarity to ideas about the structure and function of the classical ether that for a time survived the relativity and quantum revolutions. As mentioned in Chap. 2, Lodge arrived independently (and without considering quantum theory) at a result for the energy density of the ether of the same order as the one calculated by Wiechert. The British physicist shared some of the cosmological views of Nernst and Wiechert, including the idea that matter particles generated from the potential energy of the ether might act counter-entropically and prevent the heat death of the universe. He likewise speculated that radioactivity might not be limited to processes of degeneration, but also involve regeneration of matter.

The ideas that Nernst entertained with regard to ether and zero-point energy seem to have been well known in Germany. However, they did not attract much scientific interest among mainstream physicists, who may not have found his arguments for a vacuum zero-point radiation convincing. The general attitude may rather have been the one summarized by Siegfried Valentiner (1919, p. 41), professor of physics at the

Mining Academy in Clausthal: "It is much more difficult to conceive the presence of such a zero-point energy in the vacuum filled with electrical radiation than it is to assume that the existence of the zero-point energy is a peculiarity of the [material] oscillators."

The cosmological considerations of Nernst were positively reviewed by Paul Günther (1924), a physical chemist and former student of his, and they were disseminated to wider audiences in both Europe, Russia and the United States. Nernst's use of the cosmic zero-point energy as a means to counteract the entropic heat death was occasionally noticed in the philosophical and theological debate concerning the end of the world, but he preferred to address physicists and chemists rather than philosophers and theologians. Fritz Reiche (1918, 1921, p. 33) was among the few quantum physicists who referred to Nernst's theory, which he did by briefly dealing with the "radical" claim of a zero-point radiation filling all of space. So did Richard Tolman (1920, p. 1189) in Pasadena, who commented on the ideas of Keesom, Nernst and Stern: "This 'nullpunkt energie' in the Nernst treatment is in equilibrium with radiant energy in the ether. On rise of temperature, energy is drawn not only from the surroundings but also from the reservoir of 'nullpunkt energie' and the principle of the conservation of energy becomes merely statistically true rather than true for the individual elements of the system." Tolman found Nernst's ideas of zero-point energy and energy nonconservation to be interesting, but he saw no reason to adopt them.

It is, finally, worth mentioning that Bohr was also aware of Nernst's idea of 1916 that energy may not be conserved in an absolute but only statistical sense. In an unpublished manuscript from 1917 or 1918 (Bohr 1984, p. 15), he referred to "an interesting attempt to build up a theory on this basis [which] has been made by Nernst." He was thus aware of Nernst's version of vacuum energy, but chose not to comment on it.

Apart from occasional references to Nernst's ideas of a vacuum zero-point radiation in the 1910s and 1920s, his hypothesis was effectively forgotten. Only much later, and in particular with the advent of dark energy, did it attract some renewed attention. Dennis Sciama (1978) referred briefly to Nernst's theory of 1916, as did Charles Enz in greater detail (Enz 1974). In a non-cosmological context the hypothesis reappeared in the late 1960s, when Timothy Boyer at the University of Maryland proposed a theory of electromagnetic zero-point energy that became one of the sources of the research program known as "stochastic electrodynamics."[3] As Boyer (1969) pointed out, some features of his theory had been anticipated by Nernst more than fifty years earlier.

---

[3] The zero-point energy of stochastic electrodynamics is not based in quantum mechanics, but has its origin in fluctuations of classical electromagnetic fields. Indeed, some advocates of stochastic electrodynamics see the research programme as a partial alternative to quantum mechanics, in the sense that quantum effects are attributed to the classical zero-point field (Boyer 1980).

References

Barkan, D.K.: Walther Nernst and the Transition to Modern Physical Science. Cambridge University Press, Cambridge (1999)

Bartel, H.-G., Huebener, R.P.: Walther Nernst: Pioneer of Physics and of Chemistry. World Scientific, Singapore (2007)

Bohr, N.: Niels Bohr. Collected Works. vol. 5. Stolzenburg, K. (ed.) Amsterdam (1984)

Bohr, N.: On the constitution of atoms and molecules. Philos. Mag. **26**(1–25), 476–502, 851–875 (1913)

Boyer, T.H.: A brief survey of stochastic electrodynamics. In: Foundations of Radiation Theory and Quantum Electrodynamics, pp. 49–63. Springer, New York (1980)

Boyer, T.H.: Derivation of the blackbody radiation spectrum without quantum assumptions. Phys. Rev. **182**, 1374–1383 (1969)

Bromberg, J.: The concept of particle creation before and after quantum mechanics. Hist. Stud. Phys. Sci. **7**, 161–182 (1976)

Browne, P.F.: The cosmological views of Nernst: an appraisal. Apeiron **2**, 72–78 (1995)

Debye, P.: Die Konstitution des Wasserstoff-Moleküls, pp. 1–26. Sitzungsberichte, Bayerischen Akademie der Wissenschaften München, Munich (1915)

Enz, C.P.: Is the zero-point energy real? In: Enz, C.P., Mehra, J. (eds.) Physical Reality and Mathematical Description, pp. 124–132. Reidel, Dordrecht (1974)

Gearhart, C.A.: "Astonishing successes" and "bitter disappointment": the specific heat of hydrogen in quantum theory. Arch. Hist. Exact Sci. **64**, 113–202 (2010)

Günther, P.: Die kosmologische Betrachtungen von Nernst. Zeitschrift für Angew. Chem. **37**, 454–457 (1924)

Huber, P., Jaakkola, T.: The static universe of Walther Nernst. Apeiron **2**, 53–57 (1995)

Haas, A.E.: Ist die Welt in Raum und Zeit unendlich? Archiv für Systematische Philosophie **18**, 167–184 (1912)

Kragh, H.: Cosmology between the wars: the Nernst-Macmillan alternative. J. Hist. Astron. **26**, 93–115 (1995)

Kragh, H.: Entropic Creation: Religious Contexts of Thermodynamics and Cosmology. Ashgate, Aldershot (2008)

Langevin, P., de Broglie, M. (eds.): La Théorie du Rayonnement et les Quanta: Rapports et Discussions de la Réunion Tenue à Bruxelles, du 30 Octobre au 3 Novembre 1911. Gauthier-Villars, Paris (1912)

Nernst, W.: Zur theorie der spezifischen Wärme und über die Anwendung der Lehre von den Energiequanten auf physikalisch-chemische Fragen überhaupt. Zeitschrift für Electrochemie **17**, 265–275 (1911)

Nernst, W.: Zur neueren Entwicklung der Thermodynamik. Verhandlungen der Ges. Deutscher Naturforscher und Ärtzte **1**, 100–116 (1912)

Nernst, W.: Über einen Versuch, von quantentheoretischen Betrachtungen zur Annahme stetiger Energieänderungen zurückzukehren. Verhandlungen der Deutschen Phys. Ges. **18**, 83–116 (1916)

Nernst, W.: Das Weltgebäude im Lichte der Neueren Forschung. Julius Springer, Berlin (1921)

Nernst, W.: Zum Gültigkeitsbereich der Naturgesetze. Die Naturwissenschaften **10**, 489–495 (1922)

Nernst, W.: Theoretische Chemie vom Standpunkte der Avagadroschen Regel und der Thermodynamik. Ferdinand Enke, Stuttgart (1926)

Nernst, W.: Die Strahlungstemperatur des Universums. Ann. der Phys. **32**, 44–48 (1938)

Partington, J.R.: The Nernst memorial lecture: Hermann Walther Nernst. J. Am. Chem. Soci. **75**, 2853–2872 (1953)

Peebles, P.J.E., Ratra, B.: The cosmological constant and dark energy. Rev. Mod. Phys. **75**, 559–606 (2003)

Reiche, F.: Die Quantentheorie: Ihr Ursprung ind ihre Entwicklung. Die Naturwissenschaften **6**, 213–230 (1918)

Reiche, F.: Die Quantentheorie: Ihr Ursprung und ihre Entwicklung. Julius Springer, Berlin (1921)

Sciama, D.W.: The ether transmogrified. New Sci. **77** (2 February), 298–300 (1978)

Sutherland, W.: Cathode, Lenard and Röntgen rays. Philos. Mag. **47**, 269–284 (1899)

Tolman, R.C.: The entropy of gases. J. Am. Chem. Soc. **42**, 1185–1193 (1920)

Valentiner, S.: Die Grundlagen der Quantentheorie in Elementarer Darstellung. Vieweg & Sohn, Braunschweig (1919)

Wiechert, E.: Der Äther im Weltbild der Physik. Nachrichten von der Königlichen Gesellschaft der Wissenschaften zur Göttingen. Math.-Phys. Klasse **1**, 29–70 (1921a)

Wiechert, E.: Anmerkungen zur Theorie der Gravitation und über das Schicksal der Gestirne. Vierteljahrschrift der Astron. Ges. **56**, 171–191 (1921b)

Chapter 6
The Hamburg Connection

Abstract In the mid-1920s Nernst's hypothesis of a universe filled with zero-point radiation was considered within the framework of the static cosmological model that Einstein had suggested in 1917. With this work quantum theory was introduced into relativistic cosmology for the first time, but the result was disappointing. Wolfgang Pauli and other leading physicists, including Einstein, maintained that zero-point energy could be ascribed to material systems only. When it turned up in the free electromagnetic field, it was as a mathematical artefact. Nernst's hypothesis thus appeared to be a blind alley.

Keywords Zero-point energy · Electromagnetic field · Cosmic radiation temperature · Static universe model

One person who briefly alluded to Nernst's work on zero-point energy was Otto Stern, who investigated the conditions for cosmic equilibrium between radiation and matter in a pair of paper from 1925 to 1926. At the a time a professor of experimental physics at the University of Hamburg, he was inspired by Arthur Eddington's recent theory of stellar evolution, according to which the radiation energy from the stars was the result of matter-to-radiation nuclear processes, in the form of either proton-electron annihilation or fusion of hydrogen into helium. Eddington first suggested his theory of stellar energy generation in 1920 and later expounded it in his authoritative monograph *The Internal Constitution of the Stars* (Eddington 1926). In this context Stern referred to the possibility of inverse processes in which matter was produced by radiation energy: "In order to save the world from the heat death, Nernst once proposed the hypothesis that atoms of high atomic number might spontaneously be created by the radiation in cosmos [*Weltraumstrahlung*], to which he ascribed a zero-point energy" (Stern 1925, p. 448).

Stern (1926a; translated in Stern 1926b) considered a hollow space in equilibrium, meaning that the portion of matter radiated away in unit time would equal the amount of matter formed from the radiation. Although he did not explicitly introduce a cosmological perspective, he found it "very tempting to assume that cosmic space is in this state of equilibrium." Stern's universe was a gigantic cavity filled with matter and radiation. Assuming the volume V to be fixed and the matter particles of mass m to behave like an ideal gas, he calculated the maximum entropy and in this way derived an expression for the number of particles n per unit volume in equilibrium with blackbody radiation at temperature T:

$$n = \frac{N}{V} = \frac{(2\pi mkT)^{3/2}}{h^3} \exp\left(-\frac{mc^2}{kT}\right) . \tag{6.1}$$

On account of the dominating effect of the exponential term, the concentration of particles was exceedingly small even at very high equilibrium temperatures. In addition, at a given temperature the number of protons would be much smaller than the number of electrons. According to Stern it would need a temperature of about 100 million degrees to support a particle density of one electron per cm^3, and for protons the temperature would be nearly 2000 times as great, $T \cong 10^{11}$ K. This posed a problem, for these temperatures were much larger than the 30 million K that Eddington had calculated for the interior of typical stars, and thus implied an almost totally radiation-dominated universe. The result was also incompatible with the known electro-neutrality of matter: electrons and hydrogen nuclei had to be equally abundant, or very nearly so. One possible solution was in sight, but one that Stern chose to relegate to a footnote: "If any zero-point energy is to be ascribed to the radiation (Nernst) ... [it] would lower the temperatures calculated" (Stern 1926a, p. 62; Tolman 1934, pp. 147–151).

Stern's paper triggered some further work on the subject, in particular by Wilhelm Lenz and Pascual Jordan in Germany, by Richard Tolman and Fritz Zwicky in the United States, and by Seitaro Suzuki in Japan. Of these we shall pay particular attention to Lenz's little-noticed contribution, which was the only one to refer to zero-point radiation. Lenz, a former student of Sommerfeld, had done important work in atomic and molecular theory and was at the time a professor of theoretical physics at the University of Hamburg, and thus a colleague of Stern (Fig. 6.1). His paper of 1926 was directly inspired by Stern's works and also mentioned Nernst's hypothesis of a *Weltraumstrahlung* (cosmic radiation). Whereas Stern had not considered thermodynamics in relation to a particular cosmological model, Lenz applied similar reasoning to the favoured relativistic model of the early 1920s, Einstein's closed and matter-filled universe proposed in 1917. (The other alternative available at the time, Willem de Sitter's model, was irrelevant since it contained no matter.). In Einstein's model, the radius of the universe R was determined by the total mass M according to

$$R = M \frac{\kappa}{4\pi^2} . \tag{6.2}$$

Here κ is Einstein's gravitational constant ($\kappa = 8\pi G/c^2$, where G is Newton's constant), which in Einstein's model is related to the cosmological constant Λ and the average density of matter ρ by $\kappa\rho = 2\Lambda$. The volume of the universe is given

Fig. 6.1 Wilhelm Lenz
(1888–1957)

by $V = 2\pi^2 R^3$. The first relation means that the radius of the universe grows with its mass, but Lenz (1926, p. 643) pointed out that "the radiation energy does not contribute to the expansion of the world." Of course, writing at a time when the publications of Lemaître and Hubble were still in the future, the expression "expansion of the world" (*Ausdehnung der Welt*) should not be understood in its modern meaning, but just as an increase of R with M. Lenz considered it an argument that weakened the credibility of a zero-point radiation in space:

> If one allows waves of the shortest observed wavelengths of $\lambda \cong 2 \times 10^{-11}$ cm (as in radioactive γ-rays[1])—and if this radiation, converted to material density ($u/c^2 \cong 10^6$), contributed to the curvature of the world—one would obtain a vacuum energy density of such a value that the world would not reach even to the moon.

Lenz showed that if a particle of mass m is created out of radiation, the radius and volume of the universe will increase by the quantities

$$\delta R = R\,\frac{m}{M} \qquad \text{and} \qquad \delta V = 3V\,\frac{m}{M}. \tag{6.3}$$

Thus, the radius is changeable and only determined if there is a definite equilibrium between radiation and matter energy. This implied that the conditions underlying Stern's calculations had to be changed, and Lenz concluded that at equilibrium the radiation energy of the Einstein world must be equal to its matter energy. By means of the Stefan–Boltzmann radiation law he found that the temperature of the radiation

---

[1] The text refers to "radioactive β-rays," but this is undoubtedly a misprint.

would depend on the world radius as

$$T^2 = \frac{1}{R}\sqrt{\frac{2c^2}{\kappa a}},$$ (6.4)

where $a = 7.6 \times 10^{-16}\,\mathrm{J\,m^{-3}\,K^{-4}}$ is the constant in the Stefan–Boltzmann law

$$\rho_{\mathrm{rad}} = aT^4.$$ (6.5)

Expressing R in cm, the expression can be written $T^2 \cong 10^{31}/R$. Lenz did not include a zero-point radiation energy in his calculations because of "the well-known uncertainties regarding this assumption." Arbitrarily assuming the radiation temperature to be 1 K, he was led to suggest a world radius of the order 10^{31} cm. Alternatively one might estimate the temperature from the radius, as given by the Einstein relation $R^2 = 2/\kappa\rho$. Taking from de Sitter the average density of matter in the universe to be $\rho \sim 10^{-26}$ g cm^{-3}, or $R \cong 10^{26}$ cm, Lenz arrived at the much too high space temperature 300 K. As to the question of electro-neutrality, that protons and electrons must be formed in equal numbers, he claimed to have solved Stern's problem: "It makes no difference whether an electron or a hydrogen nucleus is formed, or whether they radiate away."

This work by Stern and Lenz was reconsidered by Richard Tolman (1928) at the California Institute of Technology, who criticized some of Lenz's assumptions and derived formulae approximately agreeing with Stern's. In a slightly later paper also Fritz Zwicky, at the time Tolman's colleague in Pasadena, took up the equilibrium approach pioneered by Stern. Zwicky (1928, p. 592), concluded that Tolman's modification of Lenz's theory was "in clashing contradiction with the actual facts." As mentioned, neither Tolman nor Zwicky considered the effect of a zero-point energy. In a paper of 1927 also Pascual Jordan, then at the University of Göttingen, developed the approach followed by Stern and Lenz. Applying the new forms of quantum statistics (Bose–Einstein and Fermi–Dirac) to the case where the total number of particles varies, he re-derived Stern's equilibrium formula (Jordan 1927; Bromberg 1976, pp. 184–186). As a possible mechanism for matter-energy transformation in cosmic space Jordan mentioned proton-electron collisions of the kind

$$p^+ + 2e^- \rightarrow e^- + \gamma,$$ (6.6)

which process had recently been proposed by the two American physicists George Jauncey and Albert Hughes (1926). In 1928 Jordan succeeded Pauli as Lenz's assistant in Hamburg, but there is no indication that Lenz and Jordan discussed the problem of radiation in space during this period.

On the other hand, Stern presumably discussed the question of the gravitational effect of zero-point energy with Lenz in Hamburg, and we know that he discussed it with Pauli, who stayed in Hamburg between 1923 and 1928. As mentioned by Stern in his letter to Charles Enz quoted in Chap. 4, for a period of time Pauli opposed the concept of zero-point energy, and he continued to deny the reality of such an energy in free space. According to the recollection of Pauli's two last assistants, Enz and Armin Thellung, Pauli made an estimate of the gravitational effect of the zero-point radiation along the line of Nernst but with a cut-off of the classical electron radius $\lambda_{\min} = e^2/mc^2 \cong 10^{-13}$ cm. He is said to have come to the conclusion that the radius of the Einstein universe would then "not even reach to the moon".[2] A recalculation made by Norbert Straumann, who followed some of the last lectures of Pauli, results in a world radius of 31 km, definitely confirming Pauli's estimate.[3] Interestingly, the conclusion reported by Enz and Thellung is literally the same as given by Lenz in his 1926 paper, which may indicate that Pauli had discussed the issue with Lenz (which would have been natural) or at least that he was familiar with and many years later recalled Lenz's paper.

Within the context of the new Göttingen quantum mechanics, the first attempt to quantize the electromagnetic field was made by Pascual Jordan in the important *Dreimännerarbeit* from the fall of 1925, a work written jointly with Max Born and Werner Heisenberg. Modeling the field inside a cavity as a superposition of harmonic oscillators, Jordan assumed that in addition to what he called the "thermal energy" of the oscillators, there also had to exist a zero-point energy $\frac{1}{2}h \sum \nu_k$, where k denotes the degrees of freedom (Born et al. 1926, with English translation in Van der Waerden 1967). In this way he was able to derive the fluctuation formula for blackbody radiation that Einstein had obtained by statistical methods in 1909.

However, Jordan did not think of the field zero-point energy as physically real, among other reasons because of the infinite energy that would result from the infinitely many degrees of freedom of the field. "It is just a quantity of the calculation having no direct physical meaning," he wrote to Einstein in a letter of 15 December 1925 (Mehra and Rechenberg 1982–2000, vol. 6, p. 57): "One can define physically only the thermal energy in the case of $T = 0$." Einstein agreed, as he made clear in a letter to Ehrenfest dated 12 February 1926 (Mehra and Rechenberg 1982–2000, vol. 4, p. 276):

> I have continued to concern myself very much with the Heisenberg–Born scheme. More and more I tend to the opinion that the idea, in spite of all the admiration for it, is wrong. A zero-point energy of cavity radiation should not exist. I believe that Heisenberg, Born and Jordan's argument in favour of it (fluctuations) is feeble.

As to the infinity associated with the zero-point energy, Jordan soon found a way to get rid of it, namely by a substitution procedure which has been called "the first infinite subtraction, or renormalization, in quantum field theory" (Schweber 1994, pp. 108–112). The unphysical nature of the zero-point energy of space was spelled out in a paper he wrote jointly with Pauli (Jordan and Pauli 1928, p. 154):

---

[2] First reported in Enz and Thellung (1960, p. 842), and later in Enz (2002, p. 152) and many other places, e.g., Rugh and Zinkernagel (2002). Pauli told the story to Enz and Thellung, and also to Stern about 1950, but it is unclear when he made the calculation.

[3] Straumann (2009) checked the calculation while a student in Zurich, after having heard about the problem from Enz and Thellung. Neither Enz, Thellung, Straumann nor other authors commenting on the story seem to be aware of Lenz's paper. This paper contains no mention of Pauli and there is also no indication of the problem in Pauli's scientific correspondence from the 1920s (Pauli 1979).

> Contrary to the eigen-oscillations in a crystal lattice (where theoretical as well as empirical reasons speak to the presence of a zero-point energy), for the eigen-oscillations of the radiation no physical reality is associated with this "zero-point energy" of $\frac{1}{2}h\nu$ per degree of freedom. We are here dealing with strictly harmonic oscillators, and since this "zero-point energy" can neither be absorbed nor reflected—and that includes its energy or mass—it seems to escape any possibility for detection. For this reason it is probably simpler and more satisfying to assume that for electromagnetic fields this zero-point radiation does not exist at all.

In a review paper on the light quantum hypothesis from the same year, Jordan (1928, p. 195) repeated that he did not believe in a vacuum zero-point energy. Characterizing the quantity as a "blemish" (*Schönheitsfehler*), he emphasized that it should be regarded "more as a formal complication than a real difficulty".

A few years later, in an influential review of wave mechanics in the *Handbuch der Physik* (Handbook of Physics), Pauli restated his and Jordan's belief that the zero-point energy could be ascribed to material systems only and not to the free electromagnetic field. It would, Pauli wrote, "give rise to an infinitely large energy per unit volume ... [and] be unobservable in principle since it is neither emitted, absorbed nor scattered, hence it cannot be enclosed inside walls; and, as is evident from experience, it also does not produce a gravitational field" (Pauli 1933, p. 250; Enz 2002, pp. 150–153, 181). In agreement with this view, in his *Handbuch* article Pauli wrote the expression for the energy density in such a way that the zero-point energy disappeared. Although Pauli did not refer to Nernst, and may not have read his lengthy paper of 1916, implicitly his arguments were a refutation of Nernst's dynamically active zero-point energy ether. However, Pauli disregarded the creation of matter particles out of vacuum or ether fluctuations, which was a crucial point in Nernst's hypothesis and also appeared in Lenz's work of 1926. If the vacuum field produces matter, it is no longer gravitationally inert.

Apart from the proposals of Nernst and Lenz, the zero-point energy of quantum theory first appeared in a cosmological context in a note by Edward Condon at the University of Minnesota and Julian Mack at the University of Wisconsin. They phrased the problem in a manner Nernst might have approved of Condon and Mack (1930):

> According to quantum mechanics, a harmonic oscillator of frequency ν has a lowest energy state the energy of which is $\frac{1}{2}h\nu$. When the electromagnetic field is treated ... as an assemblage of independent harmonic oscillators, one of which is associated with each of the normal modes of vibration of the ether, this leads to the result that there is present in all space an infinite positive energy density. It is infinite because there is supposed to be no upper limit to the frequencies of possible normal modes.

The two American physicists admitted that their "cosmological conjecture" was a speculation rather than a scientific theory. At any rate, neither they nor other physicists developed it further. Yet it is of some interest that they conjectured that the infinite (positive) energy density of space might just cancel the infinite negative energy density postulated by Paul Dirac in his new and controversial theory of the electron.[4] According to Dirac, the infinite sea of negative energy states was itself unobservable but vacancies in it would appear as protons (or, in his later interpretation of 1931, as positive electrons, since 1933 known as positrons). He originally pictured the negative-energy sea as a vacuum: "A perfect vacuum is now to be considered as a region in which all the states of negative energy and none of those of positive energy are occupied" (Dirac 1930, p. 606). Although Dirac's extravagant imagery of a sea of negative-energy electrons was soon abandoned, it followed from the new relativistic quantum field theory that the vacuum is far from a simple entity. Within a few years it led Dirac and other physicists to characterize the vacuum as a polarizable medium.

In this context it should be pointed out that Dirac had considered the electromagnetic vacuum already in an important paper of 1927 in which he pictured the vacuum as an infinite number of unobservable photons with zero energy and momentum. Although unphysical in this state, a photon might spontaneously jump to a positive-energy state "so that it appears to have been created" (Dirac 1927, p. 261). Any number of photons might be created in this way, and so "we must suppose that there are an infinite number of light-quanta in the zero-state." The similarity in imagery to his later hole theory of electrons is striking. There is little doubt that Dirac's radiation theory of 1927 served as an inspiration for his 1931 theory that led to the concept of antiparticles (Kragh 1990, p. 96).

References

Born, M., Heisenberg, W., Jordan, P.: Zur Quantenmechanik II. Zeitschrift für Phys. **35**, 557–615 (1926)

Bromberg, J.: The concept of particle creation before and after quantum mechanics. Hist. Stud. Phys. Sci. **7**, 161–182 (1976)

Condon, E.U., Mack, J.E.: A cosmological conjecture. Nature **125**, 455 (1930)

Dirac, P.: The quantum theory of the emission and absorption of radiation. Proc. R. Soci. A **114**, 243–265 (1927)

Dirac, P.: The proton. Nature **126**, 605–606 (1930)

Eddington, A.S.: The Internal Constitution of the Stars. Cambridge University Press, Cambridge (1926)

Enz, C.P., Thellung, A.: Nullpunktsenergie und Anordnung nicht vertauschbarer Faktoren im Hamiltonoperator. Helv. Phys. Acta **33**, 839–848 (1960)

Enz, C.P.: No Time to Be Brief: A Scientific Biography of Wolfgang Pauli. Oxford University Press, Oxford (2002)

Jauncey, G., Hughes, A.L.: Radiation and the disintegration and aggregation of atoms. Proc. Natl. Acad. Sci. **12**, 169–173 (1926)

Jordan, P.: Über die thermodynamische Gleichgewichtskonzentration der kosmischen Materie. Zeitschrift für Phys. **41**, 711–717 (1927)

Jordan, P., Pauli, W.: Zur Quantenelektrodynamik ladungsfreier Felder. Zeitschrift für Phys. **47**, 151–173 (1928)

---

[4] For the historical context of Dirac's "hole" theory of the electron, see Kragh (1990, pp. 88–105); for technical and philosophical aspects, Saunders (1991).

Jordan, P.: Die Lichtquantenhypothese. Ergebnisse der Exacten Naturwissenschaften **7**, 158–208 (1928)

Kragh, H.: Dirac: A Scientific Biography. Cambridge University Press, Cambridge (1990)

Lenz, W.: Das Gleichgewicht von Materie und Strahlung in Einsteins geschlossener Welt. Phys. Z. **27**, 642–645 (1926)

Mehra, J., Rechenberg, H.: The Historical Development of Quantum Theory, vol. 6. Springer, New York (1982–2000)

Pauli, W.: Die allgemeinen Prinzipien der Wellenmechanik. In: Geiger, H., Scheel, G. (eds.) Handbuch der Physik part 1, vol. 24, pp. 83–272. Springer, Berlin (1933)

Pauli, W.: Wolfgang Pauli. Wissenschaftlicher Briefwechsel, vol. 1. Hermann, A. (ed.) et al. Springer, New York (1979)

Rugh, S.E., Zinkernagel, H.: The quantum vacuum and the cosmological constant problem. Stud. Hist. Philos. Mod. Phys. **33**, 663–705 (2002)

Saunders, S.: The negative-energy sea. In: Saunders, S., Brown, H.R. (eds.) The Philosophy of Vacuum, pp. 65–110. Clarendon Press, Oxford (1991)

Schweber, S.S.: QED and the Men Who Made It: Dyson, Feynman, Schwinger, and Tomonaga. Princeton University Press, Princeton (1994)

Stern, O.: Über das Gleichgewicht zwischen Materie und Strahlung. Zeitschrift für Electrochemie und Angew. Phys. Chem. **31**, 448–449 (1925)

Stern, O.: Über die Umwandlung von Atomen in Strahlung. Zeitschrift für Phys. Chem. **120**, 60–62 (1926a)

Stern, O.: Transformation of atoms into radiation. Trans. Faraday Soci. **21**, 477–478 (1926b)

Straumann, N.: Wolfgang Pauli and modern physics. Space Sci. Rev. **148**, 25–36 (2009)

Tolman, R.C.: On the equilibrium between radiation and matter in Einstein's closed universe. Proc. Natl. Acad. Sci. **14**, 353–356 (1928)

Tolman, R.C.: Relativity, Thermodynamics and Cosmology. Oxford University Press, Oxford (1934)

Van der Waerden, B.L. (ed.): Sources of Quantum Mechanics. Dover Publications, New York (1967)

Zwicky, F.: On the thermodynamic equilibrium in the universe. Proc. Natl. Acad. Sci. **14**, 592–597 (1928)

Chapter 7
The Cosmological Constant

Abstract The cosmological constant introduced by Einstein in 1917 was controversial, widely regarded as both promising and troubling. Early cosmologists within the new paradigm of general relativity, such as Einstein and Willem de Sitter, vaguely realized that the field equations implied a vacuum energy and pressure. The connection to the cosmological constant was only made explicit by Georges Lemaître in an address of 1933, when he suggested a vacuum energy density corresponding to about 10^{-27} g cm^{-3}. His suggestion was ignored for more than three decades. While Lemaître associated the cosmological constant with a vacuum energy, he did not consider its connection to quantum theory.

Keywords Cosmological constant · Expanding space · Vacuum energy density · Albert Einstein · Georges Lemaître

Classical ethers of the type considered by Nernst and Wiechert were not the only ones discussed in the years around 1920. Surprisingly, on the face of it, Einstein too began to speak of physical space, as described by the metrical tensor $g_{\mu\nu}$ in his general theory of relativity, as an "ether" (Kostro 2000; Overduin and Fahr 2001). In an address in Leiden in 1920 he stressed that "empty space" is not empty in the sense of having no physical properties. Quite the contrary, for he considered space to be indistinguishable from the gravitational field, which might be thought of as a non-absolute ether (Einstein 1983, pp. 3–24):

> According to the general theory of relativity, space is endowed with physical qualities; in this sense, therefore, there exists an ether. According to the general theory of relativity, space without ether is unthinkable; for in such space there not only would be no propagation of light, but also no possibility of existence for standards of space and time (measuring-rods and clocks), nor therefore any space-time intervals in the physical sense.

This is a formulation remarkably similar to Aristotle's arguments against a vacuum, as described in Chap. 1. Einstein was of the conviction that the concepts of space and ether had merged, and that the space-ether was primary relative to both matter and electromagnetic fields. Space, he said in a lecture at the University of Nottingham of 1930, "has in the last few decades swallowed ether and time and also seems about to swallow the field and the corpuscles, so that it remains the sole medium of reality" (Kostro 2000, p. 124). Although quite differently justified, Einstein's ether had the feature in common with Nernst's version that it was physically active, indeed the source of all physical activity. However, Einstein never spoke of its activity as derived from a vacuum energy, as Nernst did, and he also did not relate its physical activity to the cosmological constant, as later physicists would do.

Suggestions of a connection between vacuum energy and the cosmological constant were absent until the 1930s, although a few physicists, among them Einstein and Hermann Weyl, considered the physical meaning of the constant. Einstein justified his introduction of the cosmological term $\Lambda g_{\mu\nu}$ not only as a means of keeping the universe in a static state, but also as a means of avoiding a cosmic negative pressure. As he wrote to his friend Michele Besso on 9 March 1917, "experience teaches us that the energy density does not become negative" (Einstein 1998, p. 406).[1] At the time Einstein took the "world radius" or present value of the cosmological scale factor R to be only of order $\sim 10^7$ light years, based on a much too large estimate of $\rho = 10^{-22}\,\mathrm{g\,cm^{-3}}$ for the matter density of the universe. Nearly thirty years later, in his book *The Meaning of Relativity* (Einstein 1945, p. 111), he elaborated: "The objection to this solution [the spatially finite, uniform world model] is that one has to introduce a negative pressure, for which there exists no physical justification. In order to make that solution possible I originally introduced a new member into the field equation instead of the above mentioned pressure." It follows from the cosmological field equations, including a pressure term p, that

$$R^2 = (\Lambda - \kappa p)^{-1} . \tag{7.1}$$

Without a cosmological constant this gives $R^2 = (-\kappa p)^{-1}$, which, in order to be positive, requires $p < 0$. If the matter pressure is zero and $\Lambda > 0$, as Einstein originally assumed, the result is instead $R^2 = 1/\Lambda$.

In a paper of 1919, in which he first expressed dissatisfaction with the cosmological constant—said to be "greatly detrimental to the formal beauty of the theory"—Einstein (1919, 1952, pp. 189–198) reconsidered the connection between the constant and a negative pressure. However, in this case the context was not cosmological but an attempt at unification, namely, to provide a link between gravitation theory and the structure of electrical particles. Einstein considered an extended charged particle in the interior of which was a negative pressure, assumed to maintain equilibrium with the electromagnetic force. As Einstein probably knew, the hypothesis of an internal stress had been considered earlier by Henri Poincaré in the context of classical

---

[1] Contrary to what is generally assumed, Einstein did not originally introduce the cosmological constant in his 1917 paper in which he applied the general theory of relativity to the universe. In an extensive article published in *Annalen der Physik* in 1916 he briefly considered the field equations with an added term $\lambda g_{\mu\nu}$, where λ is a constant—not yet "cosmological." However, finding the term to be of no physical use he relegated the extended equations to an obscure footnote (Einstein 1916).

electron theory (Miller 1973). In 1906 Poincaré introduced a non-electromagnetic, negative pressure acting only on the inside of the electron, where it balanced the repulsive electromagnetic forces tending to make the electron explode. The so-called Poincaré stress was of the form

$$p = \frac{e^2}{8\pi R_e^4} , \qquad (7.2)$$

where R_e is the radius of the electron.

In regions where only electrical and gravitational forces were present, Einstein found that the cosmological constant could be expressed in terms of the Ricci or curvature scalar R (not to be confused with the scale factor) as $4\Lambda = R$. He was aware at the time that the cosmological constant can formally be replaced by a negative pressure $p = -\Lambda/\kappa$. Schrödinger had proposed doing just that in a short comment on the field equations in 1918 (Schrödinger 1918), prompting a critical response from Einstein (1998, p. 808) that began:

> When I wrote my description of the cosmic gravitational field, I naturally noticed, as the obvious possibility, the variant Herr Schrödinger had discussed. But I must confess that I did not consider this interpretation worthy of discussion.

The reason for his dismissal was that, whereas Λ was a constant, a dynamical quantity like pressure p would have to be derived as a function of space and time from some underlying theory, something Einstein deplored as leading "too deeply into the thicket of hypotheses" (Harvey 2012). With these prescient remarks Einstein anticipated both the promise and the pitfalls of modern dark energy (Chap. 10). However, neither Schrödinger nor Einstein explicitly entertained the idea of a vacuum energy with a corresponding negative pressure.

That only came much later, although there were no particular reasons for the delay. The interpretation of the cosmological constant as an effective vacuum energy density could have been made as early as 1917. Einstein expressed his field equations with the Λ-term belonging to the geometrical (left-hand) side of the equations, together with the curvature quantities $R_{\mu\nu}$ and R. They could just as easily have been written in the form

$$R_{\mu\nu} - \tfrac{1}{2}g_{\mu\nu}R = -\kappa T_{\mu\nu} + \Lambda g_{\mu\nu} , \qquad (7.3)$$

with the cosmological term contributing to the energy-stress tensor $T_{\mu\nu}$ on the material (right-hand) side. This vacuum contribution has the form of a perfect fluid with energy density $\rho_{vac}c^2$ associated with Λ and a corresponding negative pressure. The explicit form of this pressure term follows clearly from the Friedmann equations, which date from 1922 and became generally known after 1930 (Earman 2001, p. 192 and p. 206). From these equations, written with both the cosmological constant and a pressure term, it follows directly that

$$\rho_{vac} = \frac{\Lambda c^2}{8\pi G} \quad \text{and} \quad p_{vac} = -\frac{\Lambda c^4}{8\pi G} , \qquad (7.4)$$

and then

$$p_{\text{vac}} = -\rho_{\text{vac}} c^2 . \tag{7.5}$$

In the parlance of later cosmologists, the cosmological constant behaves like a fluid whose equation of state is characterized by the dimensionless parameter $w = p/\rho c^2 = -1$. (By contrast, $w = 0$ for pressureless dust-like matter, while $w = \frac{1}{3}$ for radiation or relativistic particles). In terms of the critical density introduced by Einstein and de Sitter in their flat-space cosmological model of 1932 and given by

$$\rho_{\text{crit}} = \frac{3H^2}{8\pi G} , \tag{7.6}$$

where H is the Hubble parameter, the vacuum energy density can be written as

$$\Omega_{\text{vac}} = \frac{\rho_{\text{vac}}}{\rho_{\text{crit}}} = \frac{\Lambda c^2}{3H^2} . \tag{7.7}$$

When the vacuum expands, the work done to expand it from volume V to $V + dV$ is negative, namely, $p\,dV = -\rho c^2 dV$. In spite of the expansion, the energy density of the vacuum remains constant (while the energy increases). The increase in internal energy U with expansion is best understood as a kind of inherent tension or "springiness" in the vacuum itself. (By analogy, consider a spring with $dU = kx\,dx$ whose pressure is also negative since $dU = -p\,dV$.)

The Λ-energy is sometimes described as a form of "anti-gravity" because the source of the gravitational interaction within general relativity is determined by the combination $\rho + 3p/c^2$. The pressure term can usually be neglected, but for a vacuum fluid we have

$$\rho_{\text{vac}} + \frac{3p_{\text{vac}}}{c^2} = \rho_{\text{vac}} - 3\rho_{\text{vac}} = -2\rho_{\text{vac}} . \tag{7.8}$$

But while this is negative, it should not be taken to mean that two different regions of vacuum energy, for instance, would repel each other gravitationally. Strictly speaking, gravitation in general relativity is neither an attractive nor repulsive "force" between objects; rather, it is a manifestation of the curvature of spacetime in the vicinity of those objects. When $\rho + 3p/c^2$ is positive, as for normal matter, that curvature is such that the expansion of space decreases with time. When $\rho + 3p/c^2$ is negative, as for dark energy, it increases instead. This acceleration is however only apparent on cosmological scales. It is not, for instance, measurable on the scale of the solar system (Rindler 1969).

Einstein was well aware that the cosmological constant can be considered a measure of the energy density of vacuum also on a cosmic scale. In an unpublished manuscript of early 1931 he investigated an expanding, constant-density model of the universe (O'Raifeartaigh et al. 2014). The formation of new matter, he wrote, might be due to the Λ term, for "space itself is not empty of energy." However, he concluded that this early attempt at a steady state universe was a mistake and a few months later he discarded the cosmological constant. Einstein is often quoted as saying that his original introduction of this term was the "biggest blunder" of his life. This phrase does not appear anywhere in his writings, but was attributed to him by George Gamow in his autobiography (Gamow 1970). If Einstein did say such a thing, there is not enough evidence to decide whether it amounted to a repudiation of the very idea of a cosmological constant—or a belated recognition that, by introducing Λ to keep the universe in a static state, he had missed an opportunity for what would surely have been one of the greatest *triumphs* of his life: the prediction of cosmic expansion. There is some support for the latter interpretation in an appendix to the second edition of *The Meaning of Relativity*, where Einstein wrote that "If Hubble's expansion had been discovered at the time of the creation of the general theory of relativity, the cosmologic member would never have been introduced" (Einstein 1945, p. 127).

The Dutch astronomer and cosmologist Willem de Sitter learned of the expanding universe in the early months of 1930. Although he knew that the expansion did not require a positive cosmological constant—there are expanding models with $\Lambda = 0$— he believed that the constant was in fact responsible for the expansion of space. "What is it then that causes the expansion?" he asked in a popular article (De Sitter 1931, pp. 9–10; Fig. 7.1). His answer was that "the *lambda* does it":

> It is the presence of *lambda*, the "cosmological constant" of Einstein, in the equations that not only closes up the universe, ... but also provides the possibility of its changing its size. Why it expands and does not shrink, we do not know. ... The expansion depends on the *lambda* alone. To some it may sound unsatisfactory that we are not able to point out the mechanism by which the *lambda* contrives to do it. But there it is, we cannot go beyond the mathematical equations, and ... the behavior of *lambda* is not more strange or mysterious than that of the constant of gravitation *kappa*, to say nothing of the quantum-constant h, or the velocity of light c.

By the early 1930s it was well known that the effect of Λ is equivalent to a negative pressure, which appears in some of the early reviews of the theory of the expanding universe (e.g., Zaycoff 1932 and Maneff 1932). But the equations given above, relating the energy and pressure to the cosmological constant, were only explicitly stated in 1933, when the Belgian pioneer cosmologist (and father of the big bang) Georges Lemaître spent a period as guest professor at the Catholic University of America in Washington D.C. (Fig. 7.2). In a talk given to the U.S. National Academy of Sciences on 20 November 1933, he began by noting that, with a mean density of matter $\rho \cong 10^{-30}\,\mathrm{g\,cm^{-3}}$, "If all the atoms of the stars were equally distributed through space there would be about one atom per cubic yard, or the total energy would be that of an equilibrium radiation at

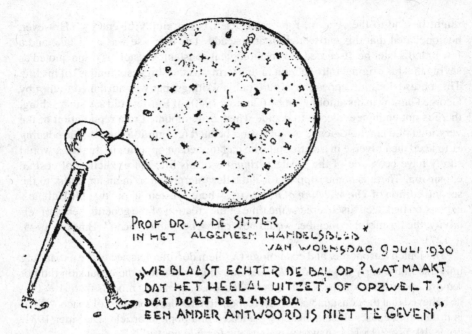

PROF DR. W DE SITTER
IN HET ALGEMEEN HANDELSBLAD *
            VAN WOENSDAG 9 JULI 1930

„WIE BLAAST ECHTER DE BAL OD ? WAT MAAKT
DAT HET HEELAL UITZET', OF OPZWELT ?
DAT DOET DE LAMBDA
EEN ANDER ANTWOORD IS NIET TE GEVEN"

Fig. 7.1 Sketch of Willem de Sitter (drawn as a "λ") in the *Algemeen Handelsblad* of 9 July 1930, as reproduced in Peebles (1993) (p. 81). De Sitter says: "What, however, blows up the ball? What makes the universe expand or swell up? That is done by lambda. No other answer can be given."

the temperature of liquid hydrogen," that is, $T \cong 20\,\mathrm{K}$ (Lemaître 1934a, p. 12).[2] A few lines later he offered the following interpretation:

> Everything happens as though the energy *in vacuo* would be different from zero. In order that absolute motion, i.e., motion relative to vacuum, may not be detected, we must associate a pressure $p = -\rho c^2$ to the density of energy ρc^2 of vacuum. This is essentially the meaning of the cosmical constant λ which corresponds to a negative density of vacuum according to $\rho_0 = \lambda c^2 / 4\pi G \cong 10^{-27}$ g/cm^3.

Notice that Lemaître's denominator was $4\pi G$ rather than $8\pi G$, and also that he took $\Lambda > 0$ to correspond to a negative ρ (Earman 2001). The negative density (and positive pressure) was not a slip of the pen, as we learn from a slightly later paper, where Lemaître (1934b) said about the cosmological constant that it "may be regarded as equivalent to a density, of negative sign, and ... accompanied with a positive pressure." While Lemaître thus offered a physical interpretation of the cosmological constant as a vacuum energy density, he did not connect his interpretation with the zero-point energy of space or otherwise relate it to quantum physics. That a

---

[2] Luminet (2007) calls this "a first intuition of a cosmic microwave background as a fossil radiation from the primeval atom," which we think is a misinterpretation. Lemaître did believe in a fossil radiation from the big bang, but he erroneously identified it with the cosmic rays made up of protons and other charged particles.

Fig. 7.2 Georges Lemaître (*right*) with Albert Einstein at the California Institute of Technology in December 1932

connection of this kind might exist seems to have been vaguely suspected by Weyl, who in a letter to Einstein of 3 February 1927 wrote: "All the properties that I had so far attributed to matter by means of Λ are now to be taken over by quantum mechanics" (Kerzberg 1989, p. 334). Alas, he did not elaborate.

Fig. 7.3 Matvei Petrovich
Bronstein (1906–1938)

Lemaître remained faithful to the idea of the cosmological constant as a form
of vacuum energy throughout his life. In part inspired by Arthur Eddington (1936),
according to whom the constant was a manifestation of the quantum nature of the
universe, he returned a few times to the subject, yet without attempting to clarify
the quantum connection and without endorsing Eddington's unorthodox theory of
cosmophysics. For example, in an address given to the 1958 Solvay conference, the
theme of which was gravitation, astrophysics and cosmology, Lemaître (1958, p. 15)
stated that, "If some extension of relativity towards a broader field, such as quantum
theory, has to be achieved the superfluous λ term shall be very much welcomed."
But instead of following up the idea, he merely remarked: "In the meantime, there
is nothing to do than to use the cosmical term in astronomical applications."

Lemaître's insight attracted little attention and failed to inspire new work related
to the strange form of vacuum energy and negative pressure. There seems to have
been no mention of it through the 1930s and 1940s, possibly because cosmologists
found Lemaître's interpretation unsurprising, or because the cosmological constant
continued to be seen as a primarily geometrical quantity within the classical theory of
general relativity, rather than a parametrization of the energy density of the vacuum
with potential relevance to quantum theory.[3] Without referring to Lemaître's paper,
Eddington (1939) argued that the cosmological constant arose if one took emptiness

---

[3] According to the Web of Science database, Lemaître's 1934 paper has (until February 2014)
been cited only 37 times, with 30 citations occurring since 1998. The late attention to his work
undoubtedly reflects the recent interest in dark energy. The Web of Science only mentions a single
citing paper in the 1930s, and this paper, by George Gamow and Edward Teller, does not refer to
the cosmological constant as a form of vacuum energy. There are no recorded citations between

to be the standard zero condition for the universe. In "seeking a self-consistent definition of the zero of energy reckoning," he wrote, the cosmological constant had to enter as the difference between the true energy density and the one measured experimentally.

One person who did explore the connection between the cosmological constant and vacuum energy (though also without reference to Lemaître) was the young Russian physicist Matvei Bronstein (Fig. 7.3). He may have been attuned to this possibility by his pioneering work on the quantization of weak gravitational fields (Gorelik and Frenkel 1994). Whatever the reason, in a paper of 1933 Bronstein not only treated the cosmological constant as a new form of matter-energy, but also suggested that it might be able to exchange energy with ordinary matter (Bronstein 1933). This marked the first suggestion that the cosmological constant might, in fact, *not be constant*, a suggestion which at the time Bronstein made "merely for the sake of generality." Much later this idea would be rediscovered and rechristened as "quintessence" (see Chap. 10).

For Bronstein, a time-varying Λ allowed him to connect cosmology to the thermodynamical arrow of time: the continual decrease in the value of Λ was tied to the fact that the universe is expanding rather than contracting. This was however a bold step, because it violated the principle of energy conservation. With $\Lambda \neq$ constant, the vacuum energy density ρ_{vac} would also no longer be constant, and the energy-conserving equation $dE + pdV = 0$ would no longer hold on cosmic scales (since $E = \rho_{vac}c^2 V$ and $p = -\rho_{vac}c^2$). Instead, Bronstein obtained the equation

$$\frac{dE}{dt} + p\frac{dV}{dt} = -\frac{\pi^2 c^2 R^3}{\kappa}\frac{d\Lambda}{dt}. \tag{7.9}$$

Far from seeing this as a fatal flaw, however, Bronstein saw it as an opportunity to connect his ideas to Bohr's suggestion at the time that energy might not be conserved in nuclear and stellar physics (Bohr 1932). Any unexplained excess radiant energy generated in stellar nuclei, he noted at the end of his paper, might be "formally equivalent to the introduction of a new form of energy connected with the λ-field which compensates Bohr's nonconservation." While radical at the time, Bronstein's insistence that the Λ term be taken seriously as a form of energy, and one that could in principle interact with the rest of physics, proved prescient, as we will see in Chap. 10.

References

Bohr, N. 1932. Atomic stability and conservation laws. In: Atti del Convegno di Fisica Nucleare, pp. 119–130. Reale Accademia d'Italia, Rome. Reprinted in Peierls, R.: (ed.) Niels Bohr Collected Works, vol. 9. North-Holland, Amsterdam (1986)

1949 and 1984. It should however be mentioned that the Web of Science is notoriously unreliable with respect to the older literature.

Bronstein, M.: On the expanding universe. Physikalische Zeitschrift der Sowjetunion **3**, 73–82 (1933)

Earman, J.: Lambda: the constant that refuses to die. Arch. Hist. Exact Sci. **55**, 189–220 (2001)

Eddington, A.S.: Relativity Theory of Protons and Electrons. Cambridge University Press, Cambridge (1936)

Eddington, A.S.: The cosmological controversy. Sci. Prog. **34**, 225–236 (1939)

Einstein, A.: Die Grundlagen der allgemeinen Relativitätstheorie. Annalen der Physik. **49**, 769–822. Reprinted in The Collected Papers of Albert Einstein, Vol. 6, Ed. A. J. Kox et al. (1996) and translated in A. Einstein et al. 1923/1952. The Principle of Relativity. Dover Publications, New York

Einstein, A.: The Collected Papers of Albert Einstein. Schulmann, R. et al. (eds.) Vol. 8A, Princeton University Press, Princeton (1998)

Einstein, A.: Spielen Gravitationsfelder im Aufbau der materiellen Elementarteilchen eine wesentliche Rolle?. Sitzungsberichte, Preussische Akademie der Wissenschaften, Berlin (1919)

Einstein, A.: The Meaning of Relativity. Princeton University Press, Princeton (1945)

Einstein, A., et al.: The Principle of Relativity. Dover Publications, New York (1952)

Einstein, A.: Sidelights on Relativity. Dover Publications, New York (1983)

Gamow, G.: My World Line. Viking Press, New York (1970)

Gorelik, G.E., Frenkel, V.Ya.: Matvei Petrovich Bronstein and Soviet Theoretical Physics in the Thirties. Birkhäuser, Basel (1994)

Harvey, A.: How Einstein discovered dark energy. ArXiv:1211.6338 [physics.hist-ph] (2012)

Kerzberg, P.: The Invented Universe: The Einstein-De Sitter Controversy (1916–1917) and the Rise of Relativistic Cosmology. Clarendon Press, Oxford (1989)

Kostro, L.: Einstein and the Ether. Apeiron, Montreal (2000)

Lemaître, G.: The primeval atom hypothesis and the problem of the clusters of galaxies. In: Stoops, R. (ed.) La Structure et l'Évolution de l'Univers, pp. 1–32. Coudenberg (1958)

Lemaître, G.: Evolution of the expanding universe. Proc. Natl. Acad. Sci. **20**, 12–17 (1934a)

Lemaître, G.: Evolution in the expanding universe. Nature **133**, 654 (1934b)

Luminet, J.-P.: The rise of big bang models, from myth to theory and observations. ArXiv:0704.3579 [astro-ph] (2007)

Maneff, G.: Über das kosmologische Problem der Relativitätstheorie. Zeitschrift für Astrophysik **4**, 231–240 (1932)

Miller, A.I.: A study of Henri Poincaré's "Sur la dynamique de l'électron". Arch. Hist. Exact Sci. **10**, 207–328 (1973)

O'Raifeartaigh, C., McCann, B., Nahm, W., Mitton, S.: Einstein's steady-state model of the universe. ArXiv:1402.0132 [physics.hist-ph] (2014)

Overduin, J., Fahr, H.-J.: Matter, spacetime and the vacuum. Naturwissenschaften **88**, 491–503 (2001)

Peebles, P.J.E.: Principles of Physical Cosmology. Princeton University Press, Princeton (1993)

Rindler, W.: Essential Relativity: Special, General and Cosmological. Van Nostrand Reinhold, New York (1969)

Schrödinger, E.: Über ein Lösungssystem der allgemein kovarianten Gravitationsgleichungen. Physikalische Zeitschrift **19**, 20–22 (1918)

De Sitter, W.: The expanding universe. Scientia **49**, 1–10 (1931)

Zaycoff, R.: Zur relativistichen Kosmogonie. Zeitschrift für Astrophysik **6**, 128–197 (1932)

Chapter 8
From Casimir to Zel'dovich

Abstract The Casimir effect predicted in 1948 was not initially seen as relevant to cosmology. For a long time the cosmological constant and the quantum mechanics of the vacuum lived separate lives. The situation only changed in the late 1960s. Inspired by a brief revival of interest in cosmological models with a positive cosmological constant, in 1968 Yakov Zel'dovich pointed out the significance of the constant in the context of quantum field theory. He also formulated the first version of what would be known as the cosmological-constant problem. With Zel'dovich's work two historical strands were finally joined: the quantum vacuum and the energy density related to the cosmological constant.

Keywords Casimir effect · Vacuum energy · Cosmological constant · Quantum fluctuations · Yakov Zel'dovich

The question of zero-point energy and related fluctuations in a pure electromagnetic field, or some other quantum field, remained unanswered for a long time. The present consensus view is that zero-point vacuum energies and fluctuations are indeed real, and that this was first evidenced by the so-called Casimir effect going back to 1948. The originator of this much-discussed effect, the Dutch physicist Hendrik Casimir (Fig. 8.1), was a former assistant of Bohr and Pauli and at the time co-director of the research laboratory of the Philips Company in Eindhoven. What Casimir (1948) predicted from the basis of quantum field theory was that there exists a non-gravitational attractive force between two perfectly conducting parallel plates even when they are placed in a vacuum. For the force per unit area he derived the expression

$$F = -\frac{1}{240}\frac{\hbar c \pi^2}{d^4}, \qquad (8.1)$$

where d is the distance between the plates. With d measured in microns, the magnitude of this quantity works out to $1.3\,\text{mN}/d^4$.

H. S. Kragh and J. M. Overduin, *The Weight of the Vacuum*, SpringerBriefs in Physics, DOI: 10.1007/978-3-642-55090-4_8, © The Author(s) 2014

Fig. 8.1 Hendrik Brugt
Gerhard Casimir (1909–2000)

Historically, the discovery of the Casimir effect arose out of an attempt to understand the van der Waals (or London-van der Waals) force first appearing in the equation of state suggested by Johannes D. van der Waals in 1873 (Sparnaay 1989; Rowlinson 2002, pp. 194–196). Casimir, collaborating with Dirk Polder, originally derived the force equation without any reference to the vacuum, and it was apparently a remark from Bohr that made him look at it from a different perspective. As he recalled, Bohr "mumbled something about the zero-point energy. He gave me a simple approach" (Bordag 1998, p. 6). At the time Bohr was well aware of the vacuum zero-point energy problem, which he had mentioned in a talk in Cambridge in the summer of 1946 and also in his address to the 1948 Solvay conference. He referred to Dirac's electron theory "which implies the existence in free space of an energy density and electric density, respectively, which ... would be far too great to conform to the basis of general relativity theory" (Bohr 1947, p. 4).

According to Casimir's communication of 1948 the difference between the zero-point energies in a pure vacuum and in the presence of the plates, given by $\frac{1}{2}h(\sum \nu_{vac} - \sum \nu_{plate})$, would be measurable. Although both terms in themselves were infinite, he could extract from them a finite result which "may be interpreted as a zero-point pressure of electromagnetic waves" (he did not explicitly refer to vacuum energy). At the end of his brief paper, he pointed out that "an experimental confirmation seems not unfeasible [*sic*] and might be of a certain interest."

During the first decade after 1948 very little interest was paid to the Casimir effect, and neither Casimir nor others related it to the vacuum energy associated with the cosmological constant. Referring to the cosmological-constant problem, in an influential review Weinberg (1989, p. 3) wrote: "Perhaps surprisingly, it was a long time before particle physicists began seriously to worry about this problem, despite the demonstration in the Casimir effect of the reality of zero-point energies." The prediction received its first qualitative support from measurements made by the Philips physicist Marcus Sparnaay (1958), but it took until 1996 before a quantitative and unambiguous test of the effect was achieved (Lamoreaux 1997). Current measurements agree with theory down to a precision of 1 % (Mohideen and Roy 1998).

The Casimir effect is real and generally taken as convincing evidence for the zero-point fluctuations of the quantum vacuum. However, it is possible to derive the effect in ways that do not refer to vacuum energy—or, by extension, to the cosmological constant. It may still be too early to claim the zero-point energy of empty space to be real beyond any doubt (Rugh et al. 1999; Jaffe 2005).

Without being aware of the potential significance of the Casimir effect, in the early 1950s William McCrea at Queens College, Belfast, suggested a new version of the steady-state theory of the universe recently introduced by Fred Hoyle, Hermann Bondi, and Thomas Gold. Hoyle's formulation of the theory was based on the field equations

$$R_{\mu\nu} - \tfrac{1}{2}g_{\mu\nu}R + C_{\mu\nu} = \kappa T_{\mu\nu} , \tag{8.2}$$

where $C_{\mu\nu}$ is a symmetric "creation tensor" of non-vanishing divergence representing the continual creation of matter in the universe. Thus, in Hoyle's theory $C_{\mu\nu}$ formally replaced the cosmological term $-g_{\mu\nu}$ in Einstein's equations of 1917. As an alternative to Hoyle's creation field, McCrea introduced a negative and non-observable negative pressure of the form $p = -\rho c^2$, identical to the equation of state (7.5) for vacuum energy discussed in Chap. 7. This hypothetical pressure he described as a "zero-point stress" responsible for the creation of new matter (McCrea 1951, pp. 573–574; Kragh 1999):

> According to relativity theory, the creation process must follow from the existence of a zero-point stress in space. Now the current quantum theory of fields endows space with several "virtual" zero-point properties. If any of these can be interpreted as producing a stress, it appears that the connexion might be established. (Such a treatment would require an examination of zero-point energy as well.)

McCrea did not refer to either Lemaître's 1934 paper or the cosmological constant, although within his theory there was a formal similarity between the postulated cosmic stress and the cosmological constant. Inspired by McCrea's idea, the Polish physicist Jaroslav Pachner (1965) proposed a cyclic model of the universe in which the singularity at the bounces (i.e., the points when the cosmic scale factor $R \to 0$) was avoided by the building up of a negative vacuum pressure varying with the curvature of space. Since then, negative pressure has been a standard ingredient in cosmologies of the cyclic or "big bounce" type (Overduin et al. 2007a; Kragh 2009).

During the first decades after the end of World War II, models with a non-zero cosmological constant were held in low esteem. Lemaître continued to advocate a positive constant, as he did at the 1958 Solvay conference on astrophysics and gravitation, but most cosmologists agreed that Einstein had been right in dismissing it as a mistake. The discovery of quasars in 1963 and the discussion of whether or not their high redshifts were of cosmological origin caused some astrophysicists to reconsider Lemaître's ideas. According to his model of 1931, the closed universe expanded from a primordial state in a way governed by a cosmological constant slightly larger than the value Λ_E in Einstein's static model:

$$\Lambda = \Lambda_E(1 + \epsilon) = \frac{\kappa\rho}{2}(1 + \epsilon) . \qquad (8.3)$$

This provided the expansion with a stagnation phase in which the cosmic scale factor $R(t)$ remained nearly constant, the length of the stagnation phase depending on the value of ϵ. Such models "loiter" at redshifts z given by $1 + z = R_0/R(t)$, where R_0 is the value of $R(t)$ at the present time (i.e., at $z = 0$), and are therefore also known in the context of modern cosmology as loitering models.

In the late 1960s, observations of the farthest quasars showed a preponderance of these unusually bright objects at redshifts $z = 1.95$, a phenomenon which could most easily be explained if the universe were in fact loitering at this redshift (so that quasars were actually of varying ages, but clustered at a single redshift). Consequently, Iosif Shklovsky (1967) and Nicolai Kardashev (1967) at the Sternberg Astronomical Institute in Moscow, as well as a few Western astrophysicists, examined models with $\Lambda > 0$. None of these works referred to the interpretation of Λ as a vacuum energy density. According to Kardashev, the quasar observations could be explained by assuming a Lemaître model with $\Lambda = 4.3 \times 10^{-56}\,cm^{-2}$ or $\epsilon = 2 \times 10^{-5}$, a model which, because of the long stagnation era, had the age of 67 billion years (Fig. 8.2). The interest in models with $\Lambda > 0$ was short-lived, as they turned out not to agree with quasar measurements after all. Vahe Petrosian (1974) at Stanford University concluded that "in the absence of strong evidence in favor of Lemaître models we must once again send back the Lemaître models to the shelf until their next re-appearance."[1]

The next re-appearance in an astronomical context may have been in 1975, when James Gunn and Beatrice Tinsley at the Hale and Lick Observatories in California analyzed the consequences of a negative value for the deceleration parameter

$$q_0 = -\left(\frac{\ddot{R}}{R}\frac{1}{H^2}\right)_0 = -\left(\frac{R\ddot{R}}{\dot{R}^2}\right)_0 , \qquad (8.4)$$

as indicated by new observations. (Here R is the cosmic scale factor and H the Hubble expansion rate. The overdot denotes differentiation with respect to time,

---

[1] Indeed, the fate of the cosmological constant since 1917 has been remarkably chequered. According to a later review, "The cosmological constant Λ is an idea whose time has come ... and gone ... and come ... and so on" (Carroll et al. 1992, p. 536). And according to a still later review, "It has been alternately reviled and praised, and it has been counted out so many times, only to stage one comeback after another" (Earman 2001, p. 189).

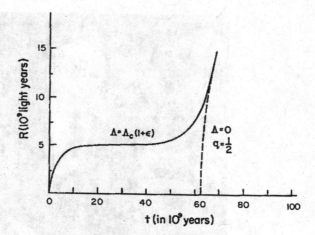

Fig. 8.2 The loitering world model of Kardashev (1967) for $H_0 = 100$ km s^{-1} Mpc^{-1} and $\epsilon = 2 \times 10^{-5}$

and the subscript "0" labels quantities evaluated at the present time; i.e., at redshift $z = 0$.) Within general relativity, $q_0 < 0$ implies an accelerated expansion that requires $\Lambda > 0$. In their article, provocatively titled "An accelerating universe," Gunn and Tinsley (1975) argued that the most plausible model was a closed Lemaître universe with $H_0 \geqslant 80$ km s^{-1} Mpc^{-1} and a density so high that no deuterium was produced during the big bang. As to the cosmological constant they admitted that its physical nature was unclear, but cited as a motivation the fact that contributions to the stress-energy tensor of the form $\Lambda g_{\mu\nu}$ "may arise quite naturally out of quantum fluctuations *in vacuo*."

The connection between the cosmological constant and the quantum mechanics of the vacuum that McCrea had vaguely anticipated was made explicit in works from the period 1965–1968. Érast Gliner, who worked at the Physico-Technical Institute in what was then Leningrad, seems to have been the first to suggest that the universe might have been begun its expansion in a vacuum-like state, an idea which eventually was developed into the inflationary scenario that will be considered in Chap. 9 (Smeenk 2005). He called the hypothetical form of vacuum-like matter a "μ-vacuum" and ascribed to it a negative pressure. In a paper of 1965, Gliner (1966, p. 381) argued that the hypothesis of a negative-pressure vacuum was not "utterly unrealistic" because attempts to describe the structure of elementary particles "would lead to the conclusion that inside the particle there must be a negative pressure which balances the electrostatic repulsion." Probably without knowing it, his remark connected to Einstein's theory of 1919 and the even earlier idea of Poincaré concerning the structure of the electron. He also seems to have been unaware of an electron model suggested "as a curiosity" by Casimir in 1953 and in which the recently predicted Casimir effect supplied the Poincaré stress; that is, the stress was explained as an effect of the zero-point energy of the electromagnetic field (Casimir 1953; Carazza and Guidetti 1986).

Yakov Zel'dovich (Fig. 8.3), a chemistry-trained nuclear physicist and cosmologist at the Institute of Applied Mathematics in Moscow, was the leading figure

Fig. 8.3 Yakov Borisovich Zel'dovich (1914–1987)

in the revival of Russian cosmology and an early contributor to the new version of big bang theory that followed in the wake of the serendipitous discovery of the cosmic microwave background radiation by Arno A. Penzias and Robert W. Wilson at Bell Labs in 1964 (see for historical review Kragh 1996). Spurred by the suggestions of Kardashev and others that quasar observations might justify models with $\Lambda > 0$, in a note of 1967 he explicitly stated what had more or less been tacitly known since the 1930s: "Corresponding to the given Λ is the concept of vacuum as a medium having a density $\rho_0 = \Lambda c^2/8\pi G = 2.5 \times 10^{-29}$ g/cm^3, an energy density $\varepsilon_0 = 2 \times 10^{-8}$ erg/cm^3, and a negative pressure (tension) $P_0 = -\varepsilon_0 = -2 \times 10^{-8}$ dyne/cm^3" (Zel'dovich 1967).

This was nothing but a restatement of what Lemaître had said more than thirty years earlier, but in a follow-up paper Zel'dovich (1968) offered a more extended discussion in which he emphasized that "A clarification of the existence and magnitude of the cosmological constant will be of tremendous fundamental significance also for the theory of elementary particles." For a long time Einstein's constant had been ignored, but now "The genie has been let out of the bottle, and it is no longer easy to force it back in. ... In our opinion, a new field of activity arises, namely the determination of Λ." Zel'dovich suggested that in principle the Λ vacuum energy might be detected in gravitation experiments of the Cavendish type—because the small density of vacuum energy in a lead sphere will contribute to its gravitational mass—but he did not refer to the Casimir effect.

The zero-point energy is calculated by representing a given field as a collection of oscillators, each oscillating with its own "mode" or frequency, and summing (or integrating) over all the modes. In principle this integral is infinite. To obtain a sensible result, Zel'dovich did what is still done today; namely, imposed a high-energy (i.e., high-frequency) cut-off, which corresponds physically to assuming that the theory of the field in question is only defined up to the cut-off energy. Arbitrarily identifying this cut-off with the rest energy of a proton (~ 1 GeV), Zel'dovich derived an electromagnetic zero-point energy ρ_{vac} of the order 10^{17} g cm^{-3}, or $\Lambda \cong 10^{-10}$ cm^{-2}, noting that this already greatly exceeded any reasonable observational bound on the value of the cosmological constant.

This was the beginning of the "cosmological-constant problem," namely, that the value of the cosmological constant as calculated on the basis of quantum zero-point fields (Λ_{zpf}) is hugely larger than bounds imposed by observation. Zel'dovich compared his calculated $\Lambda_{zpf} \cong 10^{-10}$ cm^{-2} with an observational limit $\Lambda_{obs} < 10^{-54}$ cm^{-2} or $\rho_{obs} < 5 \times 10^{-28}$ g cm^{-3}. While noting that the first value could have "nothing in common with reality," he nonetheless suggested that the cosmological constant should arise somehow from the vacuum of quantum field theory. For lack of better ideas on how to improve the agreement between theory and observation, he fell back on numerological considerations in the style of Eddington and Dirac. Based on order-of-magnitude relations between various physical constants he suggested the relationship

$$\rho_{vac} \cong \frac{GM^6 c^2}{\hbar^4} = 2 \times 10^{-11} \text{ g cm}^{-3} , \qquad (8.5)$$

where M is the mass of a proton. Another Russian physicist, M.M. Gerdov (1971), took up Zel'dovich's ideas and calculated "the mass of the virtual particle 'responsible' for the gravitation of the vacuum and the associated effect of expansion of the (Friedmann-Hubble) universe." For this particle he obtained the value $m \cong 170 m_e$, which he noted was of the same order as the mass of the muon. However, Zeldovich's numerological approach was a blind alley and did nothing to explain the observed value of the cosmological constant. (Nonetheless, it was followed by many other attempts in the same tradition.) Besides, it soon turned out that the discrepancy between theory and observation was much larger than originally estimated by Zel'dovich and his collaborators. The cosmological-constant problem remained, and has only deepened with time.

Developments in particle physics following the successful electroweak theory in the 1970s led several theorists to consider the cosmological consequences of spontaneous symmetry breaking, including the role played by the cosmological constant. Referring to the results of Zel'dovich, Andrei Linde (1974), at the time a doctoral student at the Lebedev Physics Institute in Moscow, argued that it was an "inevitable consequence" of particle physics that the vacuum energy density (and then Λ) depended on the temperature of the universe. In a hot big bang universe this implied $\Lambda = \Lambda(t)$, with almost the entire change occurring in the early universe. "The fact that Λ definitely differs from zero at a definite period of the existence of the universe makes speculations concerning a nonzero value of Λ in the present epoch more likely," he suggested. At about the same time, the problem of the impact on Λ of spontaneous symmetry breaking was discussed by Joseph Dreitlein (1974) at the University of Colorado and other authors, but there was no agreement about how broken symmetries related to the cosmological constant problem. Sidney Bludman and Malvin Ruderman (1977) thought it might be a "pseudo-problem" since even the huge vacuum energy density in the hot universe at $T \geqslant 10^{15}$ K was negligible compared to the thermal energy due to ultrarelativistic elementary particles. But the heady combination of vacuum energy and unified-field theory was soon to turn traditional ideas about the very early universe on their head.

References

Bludman, S.A., Ruderman, M.A.: Induced cosmological constant expected above the phase transition restoring the broken symmetry. Phys. Rev. Lett. **38**, 255–257 (1977)

Bohr, N.: Problems of elementary-particles physics. In: International Conference on Fundamental Particles, pp. 1–4. Cavendish Laboratory, Cambridge (1947)

Bordag, M. (ed.): The Casimir Effect 50 Years Later. World Scientific, Singapore (1998)

Carazza, B., Guidetti, G.P.: The Casimir electron model. Arch. Hist. Exact Sci. **35**, 273–279 (1986)

Carroll, S.M., Press, W.H., Turner, E.L.: The cosmological constant. Ann. Rev. Astron. Astrophys. **30**, 499–542 (1992)

Casimir, H.B.G.: Introductory remarks on quantum electrodynamics. Physica **19**, 846–849 (1953)

Dreitlein, J.: Broken symmetry and the cosmological constant. Phys. Rev. Lett. **33**, 1243–1244 (1974)

Earman, J.: Lambda: the constant that refuses to die. Arch. Hist. Exact Sci. **55**, 189–220 (2001)

Gerdov, M.M.: Concerning the nature of the cosmological constant and the mechanism of gravitation of vacuum. JETP Lett. **13**, 498–500 (1971)

Gliner, É.B.: Algebraic properties of the energy-momentum tensor and vacuum-like states of matter. Sov. Phys. JETP **22**, 378–382 (1966)

Gunn, J.E., Tinsley, B.M.: An accelerating universe. Nature **257**, 454–457 (1975)

Jaffe, R.L.: Casimir effect and the quantum vacuum. Phys. Rev. D **72**, 021301 (2005)

Kardashev, N.: Lemaître's universe and observations. Astrophys. J. **150**, L135–L139 (1967)

Kragh, H.: Steady-state cosmology and general relativity: reconciliation or conflict? In: Goenner, H. et al. (eds.) The Expanding Worlds of General Relativity, pp. 377–402. Birkhäuser, Boston (1999)

Kragh, H.: Cosmology and Controversy, pp. 52–53. Princeton University Press, Princeton (1996)

Kragh, H.: Continual fascination: the oscillating universe in modern cosmology. Sci. Context **22**, 587–612 (2009)

Lamoreaux, S.K.: Demonstration of the Casimir force in the 0.6 to 6 μm range. Phys. Rev. Lett. **78**, 5–8 (1997)

Linde, A.D.: Is the cosmological constant a constant? JETP Lett. **19**, 183–184. (The title stated in the journal, "Is the Lee constant a cosmological constant ?", is a mistranslation, as Linde pointed out in an erratum.) (1974)

McCrea, W.H.: Relativity theory and the creation of matter. Proc. Roy. Soc. A **206**, 562–575 (1951)

Mohideen, U., Roy, A.: Precision measurement of the Casimir force from 0.1 to 0.9 microns. Phys. Rev. Lett. **81**, 454904552 (1998)

Overduin, J., Blome, H.-J., Hoell, J.: Wolfgang Priester: from the big bounce to the Lambda-dominated universe. Naturwissenschaften **94**, 417–429 (2007)

Pachner, J.: An oscillating isotropic universe without singularity. Mon. Not. Roy. Astron. Soc. **131**, 173–176 (1965)

Petrosian, V.: Confrontation of Lemaître models and the cosmological constant with observations. In: Longair, M.S. (ed.) Confrontation of Cosmological Theories with Observational Data, pp. 31–46. Reidel, Dordrecht (1974)

Rowlinson, J.S.: A Scientific History of Intermolecular Forces. Cambridge University Press, Cambridge (2002)

Rugh, S.E., Zinkernagel, H., Cao, T.Y.: The Casimir effect and the interpretation of the vacuum. Stud. Hist. Phil. Mod. Phys. **3**, 111–139 (1999)

Shklovsky, I.: On the nature of the "standard" absorption spectrum of the quasi-stellar objects. Astrophys. J. **150**, L1–L3 (1967)

Smeenk, C.: False vacuum: early universe cosmology and the development of inflation. In: Kox, A.J., Eisenstaedt, J. (eds.) The Universe of General Relativity, pp. 223–257. Birkhäuser, Boston (2005)

Sparnaay, M.J.: The historical background of the Casimir effect. In: Sarlemijn, A., Sparnaay, M.J. (eds.) Physics in the Making: Essays on Developments in 20th Century Physics, pp. 235–246. North-Holland, Amsterdam (1989)

Sparnaay, M.J.: Measurements of attractive forces between flat plates. Physica **24**, 751–764 (1958)

Weinberg, S.: The cosmological constant problem. Rev. Mod. Phys. **61**, 1–23 (1989)

Zel'dovich, Ya., B.: Cosmological constant and elementary particles. JETP Lett. **6**, 316–317 (1967)

Zel'dovich, Y.B.: The cosmological constant and the theory of elementary particles. Sovjet Physics Uspekhi **11**, 381–393 (1968). Republished, with editorial introduction by V. Sahni and A. Krasinski, in Gen. Relativ. Grav. **40**: 1557–1591 (2008)

Chapter 9
Inflation and the False Vacuum

Abstract Inflation is the idea that the very early universe may have been dominated by vacuum energy, giving rise to a brief period of vastly accelerated expansion. Its roots go back to the late 1960s and the quest for singularity avoidance in the wake of the discovery of the cosmic microwave background. A tentative connection to what would later be called the horizon problem within relativistic cosmology also goes back to this period. But inflation did not gain wide recognition until high-energy particle physicists became involved. Stimulated by the successes of electroweak unification, they began to explore the implications of spontaneous symmetry breaking for cosmology, with the attendant possibility of a false vacuum state. These models implied a universe dominated by massive relic particles from the early universe. This "monopole problem" was the trigger for the explosion of interest in inflation in the 1980s. A possible explanation for flatness was noted at about the same time. Only later was it appreciated that the most compelling argument for something like inflation is its potential to link the observed large-scale structure in the present-day universe to quantum fluctuations in the earliest moments after the big bang.

Keywords Inflation · Vacuum energy · Spontaneous symmetry breaking · Cosmological constant · Large-scale structure

Cosmic inflation is a proposed period of rapid expansion driven by vacuum energy in the earliest stages of the universe (before the radiation-dominated era that produced the presently observed cosmic microwave background or CMB). As a theory, it has often been described as a product of the early 1980s, sometimes even as a natural scientific outgrowth of the era of unbridled monetary inflation, junk bonds and leveraged buyouts—"the ultimate free lunch," in the words of Alan Guth, the scientist widely credited with originating the idea. But historians, as well as key players in the story themselves, have recently painted a richer picture, showing that the stage for the inflationary era was actually set in the late 1960s, primarily by scientists in the Soviet Union (Guth 1997; Gliner 2002; Smeenk 2005; Linde 2008). Their motivation was to rescue cosmology from the spectre of the initial singularity following

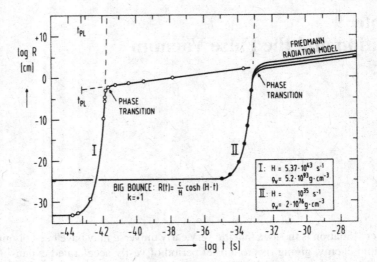

Fig. 9.1 Evolution of the cosmic scale factor R in two models of inflation in which the standard cosmological model at late times joins smoothly to an earlier nonsingular vacuum-dominated epoch. Model I (Israelit and Rosen 1989) is characterized by Planck size l_{Pl} and density ρ_{Pl} (typical of models within quantum cosmology), while Model II (Blome and Priester 1991) has $\rho_{vac} \approx 10^{-18}\rho_{Pl}$, corresponding to an energy scale of 10^{14} GeV. This latter model marked the first historical use of the term "big bounce" in cosmology (Overduin et al. 2007)

the discovery of the CMB in 1964. Although singularity avoidance is no longer cited today as a primary motivation for inflation, subsequent nonsingular models still typically incorporate a period of vacuum-driven inflationary expansion connecting the universe in its primordial state to standard "late" cosmology (Fig. 9.1).

A first step was taken by Érast Gliner (Fig. 9.2), already mentioned in connection with vacuum energy in Chap. 8. He stressed in a general way the physical reasonableness of such a medium in light of contemporary particle physics, and raised the prescient idea of a transition between a high-density state into a vacuum-like one (Gliner 1966, 1970). Similar suggestions were made independently at the same time by Sakharov (1966), though in the context of astrophysical density perturbations rather than cosmology as a whole. Zel'dovich's key contribution (Zel'dovich 1967, 1968), as already described, was to point out the link to the zero-point field of quantum field theory.[1]

It was at this time that Charles Misner at the University of Maryland revived a longstanding conundrum known as the horizon problem within standard big-bang

---

[1] Zel'dovich's chapter has been republished with commentary by Sahni and Krasiński (2008). According to a biographical article about Gliner (Chernin 2013), Zel'dovich was not enthusiastic about Gliner's vacuum model, which however found support from authorities in relativity theory such as Andrej Sakharov and Vladimir Fock.

Fig. 9.2 Érast Gliner circa 1970 in St. Petersburg (then Leningrad; reproduced from Chernin 2013)

cosmology (Misner 1969).[2] Namely, regions of space that are too widely separated to have ever been in causal contact are nevertheless observed to have precisely the same properties (for example, the same CMB radiation temperature). Causal influences propagate on scales of at most the same order as the cosmic scale factor R, whereas the horizon size $d_H = \int R \, dr$ grows much more quickly, under normal conditions at least (here r is radial coordinate distance). This can easily be seen within standard cosmology by comparing the equation for energy conservation,

$$\frac{d}{dR}\left(\rho R^3\right) = -3pR^2 \quad \rightarrow \quad \rho \propto R^{-3(1+w)}, \tag{9.1}$$

with the Friedmann equation for the scale factor (assuming for simplicity that $k = 0$):

$$\left(\frac{\dot{R}}{R}\right)^2 \propto \rho \quad \rightarrow \quad R \propto t^{2/3(1+w)}, \tag{9.2}$$

where $p = w\rho c^2$ is the equation of state. Thus cosmic time $t \propto R^{3(1+w)/2}$. Since the proper distance satisfies $ds^2 = c^2 dt^2 - R^2 dr^2 = 0$ for light rays, $R \, dr = dt$ and the size of the horizon grows as

$$d_H = \int R \, dr = \int dt \propto R^{3(1+w)/2}. \tag{9.3}$$

---

[2] Various versions of the horizon problem were known long before this, and discussed as early as 1917 by Einstein and de Sitter (Kragh 1996, pp. 234–235 and references therein).

For dust-like (pressureless) matter with $w = 0$ this means $d_H \propto R^{3/2}$, while for the hot and radiation-dominated early universe ($w = \frac{1}{3}$) it means $d_H \propto R^2$. Thus, for example, as the scale factor grows by a factor of 10^5, the horizon expands to encompass 10^{10} causally distinct regions. Since there is no reason for these regions to have the same blackbody temperature, and no way for them to interact with each other, they should show up as patches subtending about one degree in the present-day microwave sky. Why then is the actual sky so smooth?

Misner proposed to explain this isotropy by allowing causal interactions to take place via chaotic "mixmaster" oscillations in an anisotropic closed and finite world model. But Sakharov, in a preprint that only became widely known in 1982, pointed out that the same thing could be accomplished more simply with an exotic equation of state. Indeed, with a vacuum-like equation of state ($w = 0$) one sees that the horizon scale d_H can effectively be held constant as the universe grows, allowing causal influences (such as density perturbations) to communicate with each other over the entire sky. Although his proposed equation of state was not that of vacuum energy, Smeenk (2005) has remarked that this suggestion appears to be the first historical solution to the horizon problem, which would soon become one of the three pillars of inflationary cosmology.

The connection between vacuum energy and scalar fields like those required for spontaneous symmetry-breaking was made independently in 1974 by Andrei Linde at the Lebedev Physical Institute in Moscow (Linde 1974), Martinus Veltman at Utrecht University (Veltman and Martinus 1974) and Joseph Dreitlein at the University of Colorado (Dreitlein 1974). However, the three men interpreted their results in different ways. In what is arguably the forerunner of modern quintessence cosmology (Chap. 10), Linde emphasized the possibility—in his view a certainty—that such a vacuum would decay with time. Veltman focused on the huge energy density implied by theory (which had already been tentatively noted by Zel'dovich for zero-point fields in general), and expressed doubt that it could be reduced to realistic levels with any mechanism based on scalar fields, a premonition of the later "no-go" theorems of Weinberg and others (Weinberg 1989). Dreitlein suggested that the problem could be eased if the scalar were extremely light, an idea for which there is little natural motivation within existing quantum field theories, but which would later be revived by Sean Carroll (1998) and others once the reality of dark energy became established observationally.

Contributions to vacuum energy arising from the quantization of a scalar field were explored by Ernst Streeruwitz at the University of Vienna in two papers (Streeruwitz 1975a, b) that have received little attention but anticipate in some respects the work of Alexei Starobinsky a few years later. In the same year L.E. Gurevich, one of Gliner's colleagues at the Physico-Technical Institute, discussed the formation and merging of vacuum-dominated bubbles and argued that the subsequent expansion of the universe itself could be a by-product of the early vacuum-dominated phase (Gurevich 1975). Gliner, meantime, developed his more phenomenological theory with then-student Irina Dymnikova (Gliner and Dymnikova 1975). They noted that the cosmic scale factor would grow tremendously during the transition from vacuum to normal matter, but did not explore the implications.

The new scalar fields inspired by theoretical developments in high-energy particle physics thus offered mixed promise as potential solutions to both the singularity and cosmological-constant problems. However, it was soon realized by Maxim Khlopov (1978), John Preskill (1979) and others that they also introduced a *new* problem: during phase transitions the density of the universe would rapidly become overwhelmed by "defects" in the fields, somewhat akin to the boundaries that form between magnetic domains in a ferromagnet as it cools. One such defect is known as the magnetic monopole (an isolated magnetic north or south pole), so the necessity for dramatically reducing the numbers of such relics became known as the monopole problem. Of course, the severity of this problem depended on one's faith in the theory that produced it. As the British cosmologist Martin Rees later observed, "Skeptics about exotic physics might not be hugely impressed by a theoretical argument to explain the absence of particles that are themselves only hypothetical" (Rees 1998). Nevertheless, the monopole problem would shortly become the second great impetus for inflation.

A third impetus, the flatness problem, was hinted at soon thereafter in a theory developed by Alexei Starobinsky at Cambridge University between 1978 and 1979 (Starobinsky 1979, 1980). In what some have described as the "first semi-realistic model of inflationary type" (though he himself did not use that term), Starobinsky followed a conservative approach, preferring not to introduce new scalar fields into physics. Rather, he used semi-classical methods to show that quantum effects within general relativity could *mimic* the behavior of a scalar field, giving rise to an initial de Sitter-like phase that evolved through a series of oscillations into the standard radiation-dominated early universe. Commenting on this theory in 1980, he noted in passing that one could neglect the effects of spatial curvature during the transitional phase, as a consequence of the rapid expansion during the de Sitter-like phase (Starobinsky 1980). At the time, he did not consider this an important point, and did not turn the argument around or realize that the observed near-flatness of the universe could therefore be construed as support for the idea of inflation in general. To his mind, the importance of his model lay in showing that one could address Misner's horizon problem without sacrificing the time-honoured assumptions of isotropy and homogeneity in cosmology (Linde 2008).

The flatness problem is closely related to the horizon problem, and indeed any inflationary model that solves the latter solves the former as well. The argument is easily grasped. From the Friedmann equation (this time retaining the curvature constant k)

$$H^2 + \frac{kc^2}{R^2} = \frac{8\pi G}{3}\rho, \tag{9.4}$$

where $H \equiv \dot{R}/R$ is the Hubble expansion rate. Dividing both sides by $8\pi G\rho/3$ and rearranging, one obtains

$$\frac{1}{\Omega_{\text{tot}}} = 1 - \frac{\alpha}{\rho R^2}, \tag{9.5}$$

where $\Omega_{tot} = \rho_{tot}/\rho_{crit}$, $\rho_{crit} = 3H^2/8\pi G$ is the critical density, and $\alpha = 3kc^2/8\pi G$ is a constant. Recalling from Eq. (9.1) that $\rho \propto R^{-3(1+w)}$, one finds that the density of the universe evolves as

$$\Omega_{tot} = \frac{1}{1 - \alpha R^{1+3w}}. \tag{9.6}$$

The problem is that this evolves very rapidly *away* from flatness (i.e., from $\Omega_{tot} = 1$) under most conditions. With $w = 0$, for instance (dust matter) $\Omega_{tot} \propto 1/R$ as R grows large, while for $w = \frac{1}{3}$ (radiation) $\Omega_{tot} \propto 1/R^2$. How then could Ω_{tot} be so close to one in the universe we observe, without extreme fine-tuning in α? A vacuum equation of state removes this problem since with $w = -1$ the R^{-2} term in the denominator rapidly becomes insignificant and $\Omega_{tot} \to 1$ as required.[3]

Starobinsky's theory was ingenious but complex, and it was slow to catch on outside the Soviet Union. In the meantime, it was overshadowed by developments in the West, where scalar field-based inflation was discovered to solve the same problems in a simpler way. The key insights came independently to three people in 1980. Alan Guth, then a visitor at the Stanford Linear Accelerator, realized in January of that year that inflation (a term he coined) would solve both the horizon and flatness problems (Guth 1981). In his model, the necessary vacuum-like equation of state resulted from the fact that the scalar field φ was trapped in a metastable state or "false vacuum"—i.e., in a local minimum, but not the "true" or global minimum of its potential energy $V(\varphi)$. It could decay out of this state and reach the true vacuum only by quantum tunnelling, a slow process that allowed the universe to expand many times over in size during a very short time. Later that year, Demosthenes Kazanas, an astrophysicist at the Goddard Space Flight Center, also noted that exponential expansion could solve the horizon problem (Kazanas 1980). Katsuhiko Sato, then a visitor at NORDITA in Copenhagen, was the first to notice that inflation would also solve a version of the monopole problem involving defects known as domain walls (Sato 1981). Because of the exponential growth in the size of the universe, the number per unit volume of unwanted relics could be made almost as small as one liked.

As Guth himself recognized, this model of inflation suffered from a critical flaw, later termed the "graceful exit problem," in that there seemed to be no easy way to bring the period of exponential expansion to an end and join it smoothly to the hot, radiation-dominated era whose imprint we see in the CMB. The universe would have cooled tremendously as it inflated. As the scalar field tunneled slowly toward its global minimum, bubbles of true vacuum would have formed spontaneously in the sea of false vacuum. These bubbles could in principle reheat the universe, but only if they interacted with each other. Instead, calculations showed that they would

---

[3] It is possible to overstate the importance of the flatness argument. Taken literally, it would appear to suggest that any present density of $\Omega_{tot,0} \neq 1$ would imply fantastic fine-tuning in the early universe under dust-like or radiation-like conditions. In fact, however, the limit $\Omega_{tot} \to 1$ is to some extent an artifact of the definition of ρ_{crit}. When a more appropriate dimensionless parameter is used, one finds that observational constraints on it are only of order one percent, which is not excessively fine-tuned (Adler and Overduin 2005).

become isolated from each other long before inflation could solve the horizon and flatness problems.

This problem was solved in 1982 by Linde (1982), and simultaneously by Paul Steinhardt and his student Andreas Albrecht at the University of Pennsylvania (Albrecht and Steinhardt 1982). Their "new inflation" model did not involve tunnelling. Inflation could begin in a metastable false vacuum, but did not need to; it could also begin with φ in an unstable initial state on one of the "slopes" of its potential $V(\varphi)$. The only requirement was that this potential have a fairly broad plateau near $\varphi = 0$. Inflation then occurred as the scalar field—by now called the "inflaton"—evolved toward the global minimum. The equation of motion of such a field is identical to that of a ball rolling down a hill, with height given by φ and the Hubble expansion rate H playing the role of friction. If the field rolled slowly relative to H, its equation of state would be vacuum-like. Inflation could be brought to a graceful end simply by having the field reach a steeper part of its potential and fall rapidly toward its minimum. Oscillations around the minimum could then reheat the universe as required.

With this realization, inflationary model-building entered a boom period that continues unabated to this day. It was soon recognized that the most compelling argument for some form of inflation does not involve the horizon, flatness or monopole problems at all. Instead, the rapid expansion of the scale factor allows us to understand the *origin of large-scale structures* like galaxies and clusters of galaxies: they are the direct descendants of quantum fluctuations in the pre-inflationary universe. It is remarkable that the same degree of inflation that solves the first three problems also successfully connects these two scales, as was first shown for Starobinsky's theory by Viatcheslav Mukhanov and Gennady Chibisov (1981). Inflation may be relevant to other issues as well, such as explaining why the universe does not rotate on large scales (Ellis and Olive 1983).[4]

Among the most important later developments was the realization by Andrei Linde (1983) that the still rather specialized inflaton potential of the new inflationary model is not necessary, if one is willing to relax one's ideas about what came *before* inflation. Prior to the 1980s, it had been taken for granted that the universe was born in a hot, radiation-dominated state of thermal equilibrium (the primordial fireball). The inflationary era began in the middle of this early radiation-dominated period when the inflaton reached the flat portion of its potential, and ended when it dropped off. But if one made no assumptions whatsoever about the state of the universe after the big bang (beyond assuming the existence of an inflaton), then different regions could have had potentials of different shapes. Some regions would have inflated, while others would not. But inflation would certainly have kicked in *somewhere*, and we would necessarily find ourselves in one such region today. Linde termed the new scenario "chaotic inflation," for obvious reasons. This new way of thinking about

---

[4] This so-called "rotation problem" is not as pressing as the others already mentioned. The experimentally verified phenomenon of frame-dragging within general relativity suggests that the inertial compass of local observers is dynamically tied to the global distribution of matter and energy. Thus any such rotation would be unobservable (Overduin 2008).

inflation has had a pervasive effect on all of cosmology. For example, it was soon realized that the value of the cosmological constant, too, could in principle range over many orders of magnitude in different regions of the early universe. In most of those regions, its value would be incompatible with the growth of large-scale structure that is required for stars, and life, to form. Steven Weinberg used this idea in 1987 to put an "anthropic bound" on the value of the cosmological constant (Weinberg 1987).

Inflation has thus done much to make the idea of vacuum energy acceptable, in the early universe at least. The remaining objections center on two main issues: fine-tuning and testability. "Chaotic inflation" models are less fine-tuned than their predecessors, in that the shape of the potential is only weakly constrained, but all models of inflation face questions about the low level of entropy that must be assumed as an initial state (Page 1983; Penrose 1989). The extent to which the "inflationary paradigm" can ever truly be falsified by experiment also continues to be debated. Observational tests at present center on the evolution of density perturbations, as revealed by the power spectrum of CMB anisotropies. Most simple models predict that this spectrum should be random or Gaussian, with a slight "tilt" (or spectral index) reflecting a deviation from pure de Sitter-like inflation (i.e., some scales inflated slightly more than others). Both these characteristics are clearly seen in data from the Wilkinson Microwave Anisotropy Probe (WMAP) and the Planck satellite. A more distinctive signature of inflation may come with observations of CMB polarization due to perturbations in the gravitational field (i.e., primordial gravitational waves). But non-inflationary solutions to the horizon and flatness problems can also be devised that agree with the data, and it is also relatively easy to come up with inflationary models that do not. It is likely that this debate will not be settled until a single, preferred model of inflation is shown to follow from deeper some underlying theory, such as quantum gravity or a grand unified theory.

References

Adler, R.J., Overduin, J.M.: The nearly flat universe. Gen. Relativ. Gravit. **37**, 1491–1503 (2005)
Albrecht, A., Steinhardt, P.J.: Cosmology for grand unified theories with radiatively induced symmetry breaking. Phys. Rev. Lett. **48**, 1220–1223 (1982)
Blome, H.-J., Priester, W.: Big bounce in the very early universe. Astron. Astrophys. **250**, 43–49 (1991)
Carroll, S.M.: Quintessence and the rest of the world: suppressing long-range interactions. Phys. Rev. Lett. **81**, 3067–3070 (1998)
Chernin, A.D.: Why does the universe expand? Zemlya i Vselennaya. Earth universe 3(2013), 50–57 (2013)
Dreitlein, J.: Broken symmetry and the cosmological constant. Phys. Rev. Lett. **33**, 1243–1244 (1974)
Ellis, J., Olive, K.A.: Inflation can solve the rotation problem. Nature **303**, 679–681 (1983)
Gliner, É.B.: Algebraic properties of the energy-momentum tensor and vacuum-like states of matter. Sov. Phys. JETP **22**, 378–382 (1966)
Gliner, É.B.: The vacuum-like state of a medium and Friedman cosmology. Sov. Phys. Dokl. **15**, 559561 (1970)

Gliner, É.B., Dymnikova, I.G.: A nonsingular Friedmann cosmology. Sov. Astron. Lett. **1**, 9394 (1975)

Gliner, É.B.: Inflationary universe and the vacuumlike state of physical medium. Phys. Usp. **45**, 213–220 (2002)

Gurevich, L.E.: On the origin of the metagalaxy. Astrophys. Space Sci. **38**, 6778 (1975)

Guth, A.: Inationary universe: a possible solution for the horizon and flatness problems. Phys. Rev. D **23**, 347356 (1981)

Guth, A.: The Inflationary Universe. Addison-Wesley, Reading (1997)

Israelit, M., Rosen, N.: A singularity-free cosmological model in general relativity. Astrophys. J. **342**, 627–634 (1989)

Kazanas, D.: Dynamics of the universe and spontaneous symmetry breaking. Astrophys. J. Lett. **241**, L59L63 (1980)

Kragh, H.: Cosmology and Controversy. Princeton University Press, Princeton (1996)

Linde, A.D.: Is the cosmological constant a constant? JETP Lett. **19**, 183–184 (1974) (The title stated in the journal, "Is the Lee constant a cosmological constant ?", is a mistranslation, as Linde pointed out in an erratum.)

Linde, A.D.: A new inflationary universe scenario: a possible solution of the horizon, flatness, homogeneity, isotropy and primordial monopole problems. Phys. Lett. **B108**, 389–393 (1982)

Linde, A.D.: Chaotic inflation. Phys. Lett. **B129**, 177–181 (1983)

Linde, A.D.: Inflationary cosmology. Lect. Notes Phys. **738**, 1–54 (2008)

Misner, C.W.: Mixmaster universe. Phys. Rev. Lett. **22**, 10711074 (1969)

Mukhanov, V.F., Chibisov, G.V.: Quantum fluctuations and a nonsingular universe. Sov. J. Exp. Theor. Phys. Lett. **33**, 532–535 (1981)

Overduin, J., Blome, H.-J., Hoell, J.: Wolfgang Priester: from the big bounce to the Lambda-dominated universe. Naturwissenschaften **94**, 417–429 (2007)

Overduin, J.M.: The experimental verdict on spacetime from Gravity Probe B. In: Petkov, V. (ed.) Space, Time and Spacetime, pp. 25–59. Springer, Berlin (2008)

Page, D.N.: Inflation does not explain time asymmetry. Nature **304**, 39–41 (1983)

Penrose, R.: Difficulties with inflationary cosmology. Ann. N. Y. Acad. Sci. **571**, 249–264 (1989)

Preskill, J.P.: Cosmological production of superheavy magnetic monopoles. Phys. Rev. Lett. **43**, 13658 (1979)

Rees, M.: Before the Beginning. Basic Books, New York (1998)

Sahni, V., Krasiński, A.: Republication of: the cosmological constant and the theory of elementary particles (by Ya. B. Zeldovich). Gen. Relativ. Gravit. **40**, 1557–1591 (2008)

Sakharov, A.D.: The initial state of an expanding universe and the appearance of a nonuniform distribution of matter. Sov. Phys. JETP **22**, 241249 (1966)

Sato, K.: First-order phase transition of a vacuum and the expansion of the universe. Mon. Not. R. Astron. Soc. **195**, 467479 (1981)

Smeenk, C.: False vacuum: early universe cosmology and the development of inflation. In: Kox, A.J., Eisenstaedt, J. (eds.) The Universe of General Relativity, pp. 223–257. Birkhäuser, Boston (2005)

Starobinsky, A.: Spectrum of relict gravitational radiation and the early state of the universe. JETP Lett. **30**, 682685 (1979)

Starobinsky, A.: A new type of isotropic cosmological models without singularity. Phys. Lett. B **91**, 99102 (1980)

Streeruwitz, E.: Vacuum fluctuations of a quantized scalar field in a Robertson-Walker universe. Phys. Rev. D **11**, 3378–3383 (1975a)

Streeruwitz, E.: Vacuum fluctuations of a quantized scalar field in an Einstein universe. Phys. Lett. **55**, 93–96 (1975b)

Veltman, M., Martinus, J.G.: Cosmology and the Higgs mechanism. Rockefeller University Preprint, New York (1974)

Weinberg, S.: An anthropic bound on the cosmological constant. Phys. Rev. Lett. **59**, 2607–2610 (1987)

Weinberg, S.: The cosmological constant problem. Rev. Mod. Phys. **61**, 1–23 (1989)

Zel'dovich, Ya.B.: The cosmological constant and the theory of elementary particles. Sov. Phys. Usp. **11**, 381–393 (1968) (Republished, with editorial introduction by Sahni, V., Krasinski, A. in Gen. Relativ. Grav. 40 (2008): 1557–1591)

Zel'dovich, Ya.B.: Cosmological constant and elementary particles. JETP Lett. **6**, 316–317 (1967)

Zel'dovich, Y.B., Khlopov, M.Y.: On the concentration of relic magnetic monopoles in the universe. Phys. Lett. B **79**, 23941 (1978)

Chapter 10
Variable Cosmological Constants and Quintessence

Abstract Despite the success of inflation in addressing many puzzles of the early universe, theoretical cosmologists were generally reluctant to entertain the idea that vacuum energy might also dominate the universe at late times, until they were obliged to do so by astronomers. The reality of quantum zero-point fields was attested to by the Casimir effect, but this line of reasoning implied an impossibly large value for the cosmological constant. Most who were aware of the issue concluded that the latter was probably zero due to some as-yet unknown cancellation effect within quantum field theory. Some speculated that the cosmological "constant" might be a dynamical variable, perhaps associated with a scalar field like the ones proposed in models of spontaneous symmetry breaking. It might then have decayed from large primordial values to the much smaller ones compatible with present-day observational cosmology. These ideas were revived in the 1990s under the name of quintessence. Quintessence cosmology remains a very active field of research, but may raise as many new questions as it answers.

Keywords Cosmological constant · Dark energy · Quintessence cosmology · Scalar-tensor theory · General relativity

It is interesting that more theorists did not seize on the successes of inflation to propose a similar era of vacuum-driven inflation at *late times*—at least not until after this was forced on them by the detection of dark energy by observational astronomers. To understand the reasons for this reticence, we must go back to the cosmological-constant problem, the severity of which began to be appreciated only two years after Érast Gliner's proposal of a vacuum-dominated early universe.

The cosmological-constant problem is the mismatch between the tremendous vacuum energy density implied by the zero-point fields of quantum field theory and the actual density of the universe. Even without doing any measurements, we know that the latter cannot be much larger than the present *critical density*

H. S. Kragh and J. M. Overduin, *The Weight of the Vacuum*, SpringerBriefs in Physics, DOI: 10.1007/978-3-642-55090-4_10, © The Author(s) 2014

Table 10.1 Theoretical estimates of zero-point vacuum energy density and related quantities for modern quantum field theories, assuming cutoffs at the relevant characteristic energy scales

Theory	Cutoff (GeV)	Predicted value of ρ_{vac}	$\Omega_{\Lambda,0}$	Λ (cm^{-2})	R_Λ (cm)
QCD	0.3	$(0.3)^4 \hbar^{-3} c^{-5} = 10^{16}\,\mathrm{g\,cm}^{-3}$	10^{44}	10^{-12}	10^5
EW	200	$(200)^4 \hbar^{-3} c^{-5} = 10^{26}\,\mathrm{g\,cm}^{-3}$	10^{55}	1	1
GUTs	10^{19}	$(10^{19}\,\mathrm{GeV})^4 \hbar^{-3} c^{-5} = 10^{93}\,\mathrm{g\,cm}^{-3}$	10^{122}	10^{66}	10^{-34}

$\rho_{\mathrm{crit},0} = 3H_0^2/8\pi G$ for which space is flat and geometry Euclidean.[1] For densities much greater than this, the geometry of space would be noticeably spherical; and for smaller densities it would be hyperbolic. As it happens, the critical density is close to the density of matter that we can see in the universe. Thus the energy density of the vacuum, whatever its origin, cannot greatly exceed that of visible matter, which is very thinly spread indeed: the equivalent of about 0.24 protons per cubic meter.

As already noted, the problem was first explicitly brought into the open by Yakov Zel'dovich (1967, 1968). (It was present implicitly long before that, since zero-point energies are implied by quantum theory, and all energy gravitates in relativity theory, but Zel'dovich seems to have been the first to call attention to the consequences.) The level of disagreement was actually considerably worse than implied by his calculation (which arbitrarily assumed a cutoff of 1 GeV in the contributions of individual modes to the total zero-point energy). Experimental verification of the Casimir effect, while driving home the reality of the problem, also underestimated its severity insofar as it involved only zero-point contributions from the electromagnetic field. *All* fields must contribute. Repeating Zel'dovich's calculation with appropriate cutoff energies characteristic of the field theories of the current standard model of particle physics (electroweak or EW theory and quantum chromodynamics or QCD for the strong force), along with reasonable extrapolations of them (grand unified theories or GUTs that might unify EW with QCD), one finds predicted vacuum energy densities as much as *122 orders of magnitude* greater than the critical density (Table 10.1).

In this table we follow standard notation whereby densities are expressed relative to the present value of the critical density, so that $\Omega_{\Lambda,0} \equiv \rho_{vac}/\rho_{\mathrm{crit},0}$.[2] The corresponding value of the cosmological constant is

$$\Lambda = \frac{8\pi G \rho_{vac}}{c^2} = \frac{3H_0^2 \Omega_{\Lambda,0}}{c^2}. \tag{10.1}$$

---

[1] Here H_0 is the present value of the Hubble expansion rate H. Observations currently indicate that $H_0 = 74 \pm 2\,\mathrm{km\,s}^{-1}\,\mathrm{Mpc}^{-1}$ (Riess et al. 2011), implying a present critical density $\rho_{\mathrm{crit},0}$ equivalent to 6.1 proton masses per cubic meter. We follow standard notation, whereby the subscript "0" denotes any quantity evaluated at the present time.

[2] Even if Λ and ρ_{vac} are constant, $\Omega_\Lambda = \rho_{vac}/\rho_{crit}$ is not, because ρ_{crit} varies as the square of the Hubble parameter H, which depends on cosmic time.

Since Λ has dimensions of L^{-2} it defines a fundamental length scale for cosmology:

$$R_\Lambda \equiv \frac{1}{\sqrt{\Lambda}} = \frac{c}{H_0\sqrt{3\Omega_{\Lambda,0}}}. \qquad (10.2)$$

Were $\Omega_{\Lambda,0}$ actually to take values like those suggested in Table 10.1, we would see distortions of space on length scales of order R_Λ. Even taking only the predicted zero-point contributions from electroweak theory into account, space would already appear non-Euclidean on scales as small as a centimeter; whereas we know from observation that plane geometry works well out to the largest distances we can observe, of order Gpc at least. (We test this with the astronomical equivalent of adding up the angles in a triangle and seeing whether they add up to 180°.)

It is of course possible that large zero-point contributions from different field theories could cancel each other out—but this would be ludicrously improbable, the equivalent of adding two or three random numbers and obtaining a result that differed from zero only in the 122nd decimal place.

In quantum field theory, similarly large energies can usually be dealt with by a procedure known as renormalization, which essentially says that only energy differences are physically measurable, not energies themselves. This is probably one reason why it took so long for particle physicists to take the cosmological-constant problem seriously. But renormalization cannot work here, because according to general relativity *all forms of energy must gravitate*. If the zero-point energy is real, it must participate in determining the curvature of spacetime. This brings us to the heart of the problem, termed by Steven Weinberg the "one veritable crisis" remaining in physics (Weinberg 1989), and also shows why many physicists believe that a successful resolution is likely to be bound up with a successful unification of gravitation with the other fundamental interactions of physics.

There have been many attempts to resolve the cosmological-constant problem, but none has gained wide acceptance to date.[3] Historically the first, and perhaps still the most widely followed approach has been to look for a mechanism by which the density ρ_{vac} of the cosmic vacuum could be progressively "screened" or otherwise *decay* from large "primordial" values like those suggested by theory to the much smaller ones seen today. Since $\Lambda c^2 = 8\pi G\rho_{vac}$, this entails replacing Einstein's cosmological constant by a variable "cosmological term." Alternatively, as we will see in Chap. 12, the problem is reduced to explaining why the universe is of intermediate age: old enough that Λ has relaxed from primordial values like those suggested by quantum field theory to the values which we measure now, but young enough that Ω_Λ has not yet reached its asymptotic value of unity.

But how can Λ, originally introduced by Einstein as a constant of nature akin to c and G, be allowed to vary? To answer this, we go back to the field equations of general relativity (7.3) in their standard form:

---

[3] Some reviews are Adler et al. (1995), Sahni and Starobinsky (2001), Carroll (2001), Peebles and Ratra (2003) and Padmanabhan (2003, 2008).

$$R_{\mu\nu} - \tfrac{1}{2} R\, g_{\mu\nu} - \Lambda\, g_{\mu\nu} = -\frac{8\pi G}{c^4} T_{\mu\nu}. \tag{10.3}$$

Here $g_{\mu\nu}$ is the metric tensor (a generalization of Pythagoras' theorem that specifies intervals between events in curved spacetime), $R_{\mu\nu}$ is the Ricci tensor (composed of first and second derivatives of the metric), and R is the Ricci or curvature scalar (a contraction or "inner product" of $R_{\mu\nu}$ with the metric), and $T_{\mu\nu}$ is the energy-momentum tensor (containing all the properties of matter). In general there are ten equations here, because and the indices μ and ν range over 0 (time) and 1,2,3 (space), and the tensors are symmetric; but in standard cosmology there is enough additional symmetry that the number of independent equations reduces to two—one for space and one for time.

The right-hand side of Eq. (10.3) must be conserved (conservation of energy and momentum), so the same must hold true for the left-hand side. That is, the "derivatives" of both sides must vanish. In general relativity, the partial derivative ∂_μ is extended to a fully tensorial "covariant derivative" ∇_μ that reduces to the partial derivative in flat space with Cartesian coordinates. The covariant derivative of Eq. (10.3) can be written in the following form with the help of the Bianchi identities, which read $\nabla^\nu(R_{\mu\nu} - \tfrac{1}{2} R\, g_{\mu\nu}) = 0$:

$$\partial_\mu \Lambda = \frac{8\pi G}{c^4}\, \nabla^\nu T_{\mu\nu}. \tag{10.4}$$

Within Einstein's theory, it follows that $\Lambda = $ constant as long as matter and energy (as contained in $T_{\mu\nu}$) are conserved.

In variable-Λ theories, one must do one of three things: abandon matter-energy conservation, modify general relativity, or stretch the definition of what is conserved. As we have seen, the first of these routes was explored as early as 1933 by Matvei Bronstein, who sought to connect energy non-conservation with the cosmological arrow of time. Few physicists today would be willing to sacrifice energy conservation outright. Some, however, would be willing to modify general relativity, or to consider new forms of matter and energy.

Historically, these two approaches have sometimes been seen as distinct, with one being a change to the geometry of nature while the other is concerned with the material content of the universe. Einstein himself compared the left and right sides of Eq. (10.3) to the wings of a house, one made of "fine marble" and the other of "low-grade wood" (Einstein 1963). While the left-hand side followed from considerations of mathematical elegance alone, the terms on the right-hand side had to be put in "by hand." But he never wavered in his conviction that the two wings would eventually be seen as one. In the posthumous fifth edition of *The Meaning of Relativity* (Einstein 1945), he wrote:

> The present theory of relativity is based on a division of physical reality into a metric field (gravitation) on the one hand, and matter on the other. In reality, space will probably be of a uniform character, and the present theory will be valid only as a limiting case.

This vision, often referred to as the "geometrization of physics" continues to animate physicists today, although "physicalization of geometry" would be equally apt (Bergmann 1979). Nowhere is this clearer than in the relationship between vacuum energy and the cosmological constant.

Let us see how this works in one of the oldest and simplest variable-Λ theories: a modification of general relativity in which the metric tensor $g_{\mu\nu}$ is supplemented by a scalar field φ whose coupling to matter is determined by a parameter ω (not to be confused with the equation-of-state parameter w).[4] Ideas of this kind go back to Pascual Jordan (1949), Markus Fierz (1956) and Carl Brans and Robert Dicke (1961). New fields were not then proposed in particle physics as freely as they are today, and all these authors sought to associate φ with a known quantity. Various lines of argument (notably Mach's principle) pointed to an identification with Newton's gravitational "constant" such that $G \sim 1/\varphi$. By 1968 it was appreciated that Λ and ω too would depend on φ in general (Bergmann 1968). The original Brans-Dicke theory (with $\Lambda = 0$) has subsequently been extended to generalized scalar-tensor theories in which $\Lambda = \Lambda(\varphi)$ (Endō and Fukui 1977), $\Lambda = \Lambda(\varphi)$, $\omega = \omega(\varphi)$ (Barrow and Maeda 1990) and $\Lambda = \Lambda(\varphi, \psi)$, $\omega = \omega(\varphi)$ where $\psi \equiv \partial^\mu \varphi \, \partial_\mu \varphi$ (Fukui and Overduin 2002). In the last and most general of these cases, the field equations become

$$R_{\mu\nu} - \tfrac{1}{2} R \, g_{\mu\nu} + \frac{1}{\varphi} \left[\nabla_\mu (\partial_\nu \varphi) - \Box \varphi \, g_{\mu\nu} \right] + \frac{\omega(\varphi)}{\varphi^2} \left(\partial_\mu \varphi \, \partial_\nu \varphi - \frac{1}{2} \psi \, g_{\mu\nu} \right)$$

$$- \Lambda(\varphi, \psi) \, g_{\mu\nu} + 2 \frac{\partial \Lambda(\varphi, \psi)}{\partial \psi} \, \partial_\mu \varphi \, \partial_\nu \varphi = -\frac{8\pi}{\varphi c^4} T_{\mu\nu}, \tag{10.5}$$

where $\Box \varphi \equiv \nabla^\mu (\partial_\mu \varphi)$ is the D'Alembertian. These reduce to Einstein's field equations (10.3) when $\varphi = 1/G =$ constant.

If we now repeat the exercise on the previous page and take the covariant derivative of the field equations (10.5) with the Bianchi identities, we obtain a generalized version of Eq. (10.4) faced by Bronstein:

$$\partial_\mu \varphi \left\{ \frac{R}{2} + \frac{\omega(\varphi)}{2\varphi^2} \psi - \frac{\omega(\varphi)}{\varphi} \Box \varphi + \Lambda(\varphi, \psi) + \varphi \frac{\partial \Lambda(\varphi, \psi)}{\partial \varphi} - \frac{\psi}{2\varphi} \frac{d\omega(\varphi)}{d\varphi} \right.$$

$$\left. - 2\varphi \Box \varphi \frac{\partial \Lambda(\varphi, \psi)}{\partial \psi} - 2 \partial^\kappa \varphi \, \partial_\kappa \left[\varphi \frac{\partial \Lambda(\varphi, \psi)}{\partial \psi} \right] \right\} = \frac{8\pi}{c^4} \nabla^\nu T_{\mu\nu}. \tag{10.6}$$

Now energy conservation ($\nabla^\nu T_{\mu\nu} = 0$) no longer requires $\Lambda =$ const. In fact, it is generally *incompatible* with constant Λ, unless an extra condition is imposed on the terms inside the curly brackets in (10.6).

---

[4] A scalar field is, in principle, the simplest possible form of matter: it is a dynamical quantity whose value in space and time can be described by a single number. Although many scalar fields have been proposed by particle theorists, only one is known to exist in nature so far: the Higgs boson. The inflaton discussed in Chap. 9 is another scalar that is widely suspected to exist, and to be responsible for inflation in the early universe.

Similar conclusions hold for other scalar-tensor theories in which φ is no longer associated with G. As noted in Chap. 9, new scalar fields of this kind began to appear in 1974 in connection with theories of spontaneous symmetry breaking by Andrei Linde (1974), Martinus Veltman (1974) and Joseph Dreitlein (1974). In the context of the early universe, they became known as inflatons. In the context of the late (present) universe, they were studied by others such as Lawrence Ford (1975), Ernst Streeruwitz (1975a, b), Hans-Joachim Blome and Wolfgang Priester (1984). Subsequent refinements have included non-minimal couplings between φ and the curvature scalar R (Madsen 1988), and conformal rescalings of the metric tensor by functions of φ (Maeda 1989). But by far the most influential innovation was the inclusion of an explicit potential $V(\varphi)$ for the scalar field (Barr 1987; Peebles and Ratra 1988; Ratra and Peebles 1988; Wetterich 1988). In such a theory the scalar field behaves like a perfect fluid with an evolving equation-of-state parameter

$$w = \frac{\frac{1}{2}\dot{\varphi}^2 - V(\varphi)}{\frac{1}{2}\dot{\varphi}^2 + V(\varphi)} \, , \tag{10.7}$$

where $\dot{\varphi}^2$ is the field's kinetic energy (the overdot represents a derivative with respect to time). A free field with $V = 0$ behaves like "stiff matter" with $w = 1$, while one with a constant kinetic term $\frac{1}{2}\dot{\varphi}^2 \approx V(\varphi)$ mimics dust-like matter with zero pressure ($w \approx 0$). As φ "rolls down" its potential and settles at a minimum, $\dot{\varphi} \to 0$ and the equation of state goes over to that of a cosmological constant, $w \to -1$.

Cosmological scalar fields of this kind were rediscovered in the 1990s and dubbed quintessence (Caldwell et al. 1998).[5] Some versions of quintessence exhibit "tracker behaviour," in which the energy density of the scalar field remains close to the energy density of radiation during the early universe (Zlatev et al. 1999). When the universe expands to the point where radiation and matter have the same density, the quintessence field "switches gears" and begins to behave like vacuum energy. This can help to explain why the present densities of matter and dark energy are so close, a coincidence that is otherwise very difficult to understand (see Chap. 12). Other versions of quintessence include "phantom fields" whose equation-of-state parameter $w < -1$, implying that the sum of pressure and energy density is negative (Caldwell et al. 2003). Such a field would have bizarre consequences for cosmic evolution, but is not necessarily inconsistent with known physics. Indeed it has become a major goal of observational astronomy to pin down the exact value of w and its possible evolution with time. To date, all the data are consistent with $w = -1$; that is, with a cosmological term that is indeed perfectly constant (Dunkley et al. 2009). But the theoretical motivation for a dynamical cosmological term is sufficiently strong that the search continues.

In the modern approach to variable-Λ cosmology, which as we have seen goes back to Gliner (1966) and Zel'dovich (1968), all extra terms of the kind just

---

[5] This was not the first revival of this term from Greek metaphysics in a modern scientific context; it was also used to describe unusual phases of matter in lunar rock samples and cosmochemistry during NASA's *Apollo* program (Overduin et al. 2007a).

described—including Λ—are moved to the right-hand side of the field equations (10.5), leaving only the Einstein tensor $R_{\mu\nu} - \frac{1}{2} R g_{\mu\nu}$ on the "geometrical" left-hand side. The cosmological term is thus interpreted as a new form of matter, along with scalar or other new fields that may be present. Equations (10.5) become

$$R_{\mu\nu} - \frac{1}{2} R g_{\mu\nu} = -\frac{8\pi}{\varphi c^4} T_{\mu\nu,\text{eff}} + \Lambda(\varphi) g_{\mu\nu}. \qquad (10.8)$$

Here $T_{\mu\nu,\text{eff}}$ is an effective energy-momentum tensor describing the sum of ordinary matter plus whatever scalar (or other) fields have been added to the theory. For generalized scalar-tensor theories as described above, this could be written as $T_{\mu\nu,\text{eff}} \equiv T_{\mu\nu} + T_{\mu\nu,\phi}$ where $T_{\mu\nu}$ refers to ordinary matter and $T_{\mu\nu,\phi}$ to the scalar field. For the case with $\Lambda = \Lambda(\varphi)$ and $\omega = \omega(\varphi)$, for instance, the latter would be defined by Eq. (10.5) as

$$T_{\mu\nu,\phi} \equiv \frac{1}{\varphi} \left[\nabla_\mu (\partial_\nu \varphi) - \Box \varphi \, g_{\mu\nu} \right] + \frac{\omega(\varphi)}{\varphi^2} \left(\partial_\mu \varphi \, \partial_\nu \varphi - \frac{1}{2} \psi \, g_{\mu\nu} \right). \qquad (10.9)$$

The covariant derivative of the field equations (10.8) then reads

$$0 = \nabla^\nu \left[\frac{8\pi}{\varphi c^4} T_{\mu\nu,\text{eff}} - \Lambda(\varphi) g_{\mu\nu} \right]. \qquad (10.10)$$

Equation (10.10) carries the same physical content as (10.6), but is more general in form and can readily be extended to other theories. Physically, it says that energy *is* conserved in variable-Λ cosmology—where "energy" is now understood to refer to the energy of matter and other fields, *and* the energy of the vacuum as represented by Λ. In general, the latter parameter can vary as it likes, so long as the conservation equation (10.10) is satisfied.

Perhaps the first to obtain an explicit expression for a time-varying Λ-term on this basis were Makoto Endō and Takao Fukui (1977). Working in the context of a scalar-tensor theory with $\Lambda = \Lambda(\varphi)$ and $\omega = $ constant, these authors found a solution for $\varphi(t)$ such that $\Lambda \propto t^{-2}$. In their early versions of quintessence cosmology, Stephen Barr (1987) similarly discussed models in which $\Lambda \propto t^{-\ell}$ at late times, while James Peebles and Bharat Ratra (1988) developed a particularly influential theory in which $\Lambda \propto R^{-m}$ at early ones (here ℓ and m are powers).[6] There is now a rich literature on Λ-decay laws of this kind and their implications for cosmology (Overduin and Cooperstock 1998). Among other things, they appear to soften or even evade some of the observational arguments against a nonsingular "big bounce" at the beginning of the current expansionary phase of the universe, a possibility that can be ruled out on quite general grounds if $\Lambda = $ constant (Overduin 1999; Fig. 10.1).

---

[6] Known as ΦCDM, the Peebles-Ratra theory continues to be an area of active research; see for example its generalization to closed and open models by Pavlov et al. (2013).

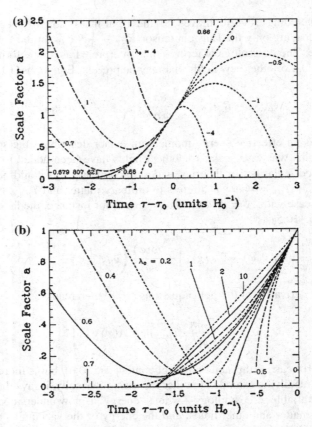

Fig. 10.1 Evolution of the cosmic scale factor $a \equiv R(t)/R_0$ as a function of time $\tau - \tau_0 \equiv H_0(t-t_0)$ (where t_0 = present), for cases in which $\Lambda \propto R^{-1}$ (**a**, *top*) and $\Lambda \propto H^2$ (**b**, *bottom*). Values of $\Omega_{\Lambda,0}$ (labeled λ_0 in the figure) are shown beside each curve. The values of $\Omega_{\Lambda,0}$ for which both of these models originate in a nonsingular "big bounce," rather than a big bang, are much lower than in standard cosmology with Λ = constant. Taken from Overduin and Cooperstock (1998)

In assessing such models, however, two caveats must be kept in mind. The first is theoretical. Insofar as the mechanisms discussed so far are entirely classical, they do not address the underlying problem. For this, one would also need to explain why net contributions to Λ from the *quantum vacuum* do not remain at their primordial level, or how they are suppressed with time. Alexander Polyakov (1982) and Stephen Adler (1982) were the first to speculate explicitly that such a suppression might come about if the "bare" cosmological term implied by quantum field theory were progressively screened by an "induced" counterterm of opposite sign, driving the effective value of $\Lambda(t)$ toward zero at late times. Many theoretical adjustment mechanisms have now been identified as potential sources of such a screening effect, beginning with a 1983 suggestion by Alexander Dolgov (1983) based on non-minimally coupled scalar fields. In most of these cases, no analytic expression is found for Λ in terms of time or other cosmological parameters; the intent is merely to demonstrate that decay (and preferably near-cancellation) of the cosmological term is possible in principle. None of these mechanisms has been widely accepted. In fact, there is a general argument due to Steven Weinberg to the effect that any successful mechanism based on scalar fields would necessarily be as fine-tuned as the cosmological-constant problem itself (Weinberg 1989). Similar concerns have been raised in the case of vector and tensor-based proposals (Dolgov 1998). More recently Richard Woodard has used a powerful theorem due to Mikhail Ostrogradsky to prove that extensions of general relativity involving additional higher-order curvature terms are equally unlikely to solve the problem: they are either equivalent to (and equally fine-tuned as) scalar quintessence, or else inherently unstable (Woodard 2007).

The second caution is empirical. Observational data place increasingly strong restrictions on the way in which Λ can vary with time. Among the most important are early-time bounds on the vacuum-energy density. The success of standard big-bang nucleosynthesis theory implies that ρ_{vac} was smaller than ρ_{rad} and ρ_{mat} during the radiation-dominated era, and large-scale structure formation could not have proceeded in the conventional way unless $\rho_{vac} < \rho_{mat}$ during the early matter-dominated era. Since $\rho_{rad} \propto R^{-4}$ and $\rho_{mat} \propto R^{-3}$, these requirements mean in practice that the vacuum-energy density must drop less steeply than R^{-3}, if it is comparable to that of matter or radiation at present. The variable-Λ term must also satisfy late-time bounds like those which have been placed on the cosmological constant by purely astrophysical means. Tests of this kind have been carried out using data on the age of the universe, structure formation, galaxy number counts, the power spectrum of temperature fluctuations in the cosmic microwave background, gravitational lensing statistics and the supernova magnitude-redshift relation.

Finally, one can ask what the vacuum decays *into*. Energy conservation requires that any decrease in the energy density of the vacuum be made up by a corresponding increase somewhere else. In quintessence theories, dark energy is transferred to the kinetic energy of the scalar field. This may have observable consequences if the scalar field is coupled strongly to ordinary matter, but is hard to constrain in general. A simpler situation is that in which the vacuum decays into known particles such as baryons, photons or neutrinos. In this case, observations of the intensity of cosmic background radiation at various wavelengths can be used to infer that a decaying vacuum component cannot make up more than about 10^{-6} times the critical density at present (Overduin et al. 1993).

References

Adler, S.L.: Einstein gravity as a symmetry-breaking effect in quantum field theory. Rev. Mod. Phys. **54**, 729–766 (1982)

Adler, R.J., Casey, B., Jacob, O.C.: Vacuum catastrophe: an elementary exposition of the cosmological constant problem. Am. J. Phys. **63**, 620–626 (1995)

Barr, S.M.: Attempt at a classical cancellation of the cosmological constant. Phys. Rev. **D36**, 1691–1700 (1987)

Barrow, J.D., Maeda, K.-I.: Extended inflationary universes. Nucl. Phys. **B341**, 294–308 (1990)

Bergmann, P.G.: Comments on the scalar-tensor theory. Int. J. Theor. Phys. **1**, 25–36 (1968)

Bergmann, P.G.: Unitary field theory, geometrization of physics or physicalization of geometry? einstein symposion berlin. Lect. Notes Phys. **100**, 84–88 (1979)

Blome, H.-J., Priester, W.: Vacuum energy in a Friedmann-Lemaître universe. Die Naturwissenschaften **71**, 528–531 (1984)

Brans, C., Dicke, R.H.: Mach's principle and a relativistic theory of gravitation. Phys. Rev. **124**, 925–935 (1961)

Caldwell, R.R., Kamionkowski, M., Weinberg, N.N.: Phantom energy: dark energy with $w \leq -1$ causes a cosmic doomsday. Phys. Rev. Lett. **91**, 071301 (2003)

Caldwell, R.R., Dave, R., Steinhardt, P.J.: Cosmological imprint of an energy component with general equation of state. Phys. Rev. Lett. **80**, 1582–1585 (1998)

Carroll, S.M.: The cosmological constant. Living Rev. Relativ. **4**, 1 (2001)

Dolgov, A.D.: In: Gibbons, G.W., Hawking, S.W., Siklos, S.T.C. (eds.) The Very Early Universe, p. 449. Cambridge University Press, Cambridge (1983)

Dolgov, A.D.: In: Sanchez, N., de Vega, H.J. (eds.) The 4th Paris Cosmology Colloquium. World Scientific, Singapore (1998)

Dreitlein, J.: Broken symmetry and the cosmological constant. Phys. Rev. Lett. **33**, 1243–1244 (1974)

Dunkley, J., et al.: Five-year Wilkinson Microwave Anisotropy Probe observations: likelihoods and parameters from the WMAP data. Astrophys. J. Suppl. Ser. **180**, 306–329 (2009)

Einstein, A.: Physics and reality. J. Franklin Inst. **221**, 349–382 (1936)

Einstein, A.: The Meaning of Relativity. Princeton University Press, Princeton (1945)

Endō, M., Fukui, T.: The cosmological term and a modified Brans-Dicke cosmology. Gen. Relativ. Gravit. **8**, 833–839 (1977)

Fierz, M.: On the physical interpretation of P. Jordan's extended theory of gravitation. Helv. Phys. Acta **29**, 128 (1956)

Ford, L.H.: Quantum vacuum energy in general relativity. Phys. Rev. **D11**, 3370–3377 (1975)

Fukui, T., Overduin, J.M.: Dynamics of a generalized cosmological scalar-tensor theory. Int. J. Mod. Phys. **D11**, 669–684 (2002)

Gliner, É.B.: Algebraic properties of the energy-momentum tensor and vacuum-like states of matter. Sov. Phys. JETP **22**, 378–382 (1966)

Jordan, P.: Formation of the stars and development of the universe. Nature **164**, 637640 (1949)

Linde, A.D.: Is the cosmological constant a constant? JETP Lett. **19**, 183–184 (1974) (The title stated in the journal, "Is the Lee constant a cosmological constant?", is a mistranslation, as Linde pointed out in an erratum.)

Madsen, M.: Scalar fields in curved spacetimes. Class. Quantum Gravity **5**, 627–639 (1988)

Maeda, K.-I.: Towards the Einstein-Hilbert action via conformal transformation. Phys. Rev. **D39**, 3159–3162 (1989)

Overduin, J.M., Wesson, P.S., Bowyer, S.: Constraints on vacuum decay from the microwave background. Astrophys. J. **404**, 1–7 (1993)

Overduin, J.M., Cooperstock, F.I.: Evolution of the scale factor with a variable cosmological term. Phys. Rev. D **58**, 043506 (1998)

Overduin, J.M.: Nonsingular models with a variable cosmological term. Astrophys. J. **517**, L1–L4 (1999)

Overduin, J., Blome, H.-J., Hoell, J.: Wolfgang Priester: from the big bounce to the Λ-dominated universe. Naturwissenschaften **94**, 417–429 (2007)

Padmanabhan, T.: Cosmological constant—the weight of the vacuum. Phys. Rep. **380**, 235–320 (2003)

Padmanabhan, T.: Dark energy and gravity. Gen. Relativ. Gravit. **40**, 529–564 (2008)

Pavlov, A., et al.: Nonflat time-variable dark energy cosmology. Phys. Rev. D **88**, 123513 (2013)

Peebles, P.J.E., Ratra, B.: Cosmology with a time-variable cosmological 'constant'. Astrophys. J. **325**, L17–L20 (1988)

Peebles, P.J.E., Ratra, B.: The cosmological constant and dark energy. Rev. Mod. Phys. **75**, 559–606 (2003)

Polyakov, A.M.: Phase transitions and the universe. Sov. Phys. Usp. **25**, 187 (1982)

Ratra, B., Peebles, P.J.E.: Cosmological consequences of a rolling homogeneous scalar field. Phys. Rev. **D37**, 3406–3427 (1988)

Riess, A.G. et al.: A 3% solution: determination of the Hubble constant with the Hubble Space Telescope and Wide Field Camera 3. Astrophys. J. **730**, 119 (18 pp) (2011)

Sahni, V., Starobinsky, A.: The case for a positive cosmological Λ-term. Int. J. Mod. Phys. A **D9**, 373–444 (2001)

Streeruwitz, E.: Vacuum fluctuations of a quantized scalar field in a Robertson-Walker universe. Phys. Rev. D **11**, 3378–3383 (1975)

Streeruwitz, E.: Vacuum fluctuations of a quantized scalar field in an Einstein universe. Phys. Lett. **55**, 93–96 (1975)

Veltman, M.J.G.: Cosmology and the Higgs Mechanism. Rockefeller University Preprint, New York (1974)

Weinberg, S.: The cosmological constant problem. Rev. Mod. Phys. **61**, 1–23 (1989)

Wetterich, C.: Cosmology and the fate of dilatation symmetry. Nucl. Phys. **B302**, 668–696 (1988)

Woodard, R.P.: Avoiding dark energy with $1/R$ modifications of gravity. The invisible universe: dark matter and dark energy. Lect. Notes Phys. **720**, 403–433 (2007)

Zel'dovich, Y.B.: Cosmological constant and elementary particles. JETP Lett. **6**, 316–317 (1967)

Zel'dovich, Y.B.: The cosmological constant and the theory of elementary particles. Sov. Phys. Usp. **11**, 381–393 (1968). Republished, with editorial introduction by Sahni, V., Krasinski, A.: Gen. Relativ. Gravit. **40**, 1557–1591 (2008)

Zlatev, I., Wang, L., Steinhardt, P.J.: Quintessence, cosmic coincidence, and the cosmological constant. Phys. Rev. Lett. **82**, 896–899 (1999)

Chapter 11
How Heavy Is the Vacuum?

Abstract Well into the 1990s, most cosmologists preferred not to speak of the cosmological constant. This attitude was justified partly by the deep theoretical uncertainty surrounding the status of vacuum energy, and partly by the degree of fine-tuning that seemed to be implied in models whose density of vacuum energy was comparable to that of matter. Nevertheless the cosmological constant was trotted out whenever some crisis arose within cosmology that could not be explained any other way. Two examples that received attention in the 1960s were the concentration of quasars within a narrow range of high redshifts, and the tension between the age of the universe implied by measurements of the Hubble expansion rate and the age of the oldest stars. In the 1980s, vacuum energy was revived again to bridge the gap between the observed low density of matter and the expectation (based on inflation) that the total density of the universe should be exactly critical. The lack of anisotropy observed in the cosmic microwave background prior to 1992 was also taken as possible evidence for a Λ term. Tentative measurements of a nonzero dark-energy density were first obtained with counts of faint galaxies and analyses of absorption lines in the Lyman-α forest, but seemed to conflict with upper limits based on the statistics of gravitational lenses.

Keywords Dark energy · Cosmological constant · Cosmic microwave background · Galaxy counts · Lyman-alpha forest

Despite the successes of inflationary cosmology at early times, and quintessence at late ones, vacuum energy tended to be taken most seriously by high-energy physicists. Cosmologists continued to view the cosmological term with suspicion, and textbooks as late as the 1990s tended either to ignore Λ-dominated models or relegate them to appendices. This attitude was to some extent justified by a line of reasoning sometimes referred to as the "Dicke coincidence argument" (Dicke 1970; Peebles 1993), which is essentially the observation that matter and vacuum energy densities evolve very differently with time, so that a universe in which they were comparable

H. S. Kragh and J. M. Overduin, *The Weight of the Vacuum*, SpringerBriefs in Physics, DOI: 10.1007/978-3-642-55090-4_11, © The Author(s) 2014

at present would be very finely tuned. As we will see, this argument still holds: we have simply learned to live with fine tuning.

The cosmological constant was, however, dusted off whenever some crisis arose that could not be explained any other way—and then just as quickly returned to the shelf when the crisis receded. As related in Chap. 8, Λ-dominated "loitering" models were twice resurrected this way in connection with quasar redshifts in the late 1960s, and with observational data on the deceleration parameter q_0 in the 1970s. Closely related was the age problem, in which the calculated age of the universe is not enough to accommodate the oldest observed stars.[1] Models with a significant vacuum energy density $\Omega_{\Lambda,0}$ at the present time avoid this contradiction because they are older than models with the same matter density $\Omega_{M,0}$, again due to the quasi-static loitering phase that occurs when $\Omega_M \approx \Omega_\Lambda$ in the past. The age problem can also be eased with a lower Hubble constant, implying a slower expansion rate at all times in the past, and hence an older universe. Interest in Λ as a solution to this particular crisis accordingly tended to come and go depending on the prevailing opinions on the value of H_0.

Some authors stressed the importance of mathematical and presumably more "timeless" reasons to take seriously a nonzero value for the cosmological term. Dimensionally, Λ defines a fundamental length scale for cosmology, $R_\Lambda = 1/\sqrt{\Lambda}$, as already remarked in Chap. 10. Arthur Eddington alluded to this fact in 1924, writing that there must be a cosmological constant because "an electron could never have decided how large it ought to be, unless there existed some length independent of itself for it to compare itself with" (Eddington 1924). In a different vein, Wolfgang Rindler at the University of Texas argued in 1969 that setting the cosmological constant to zero in Einstein's field equations would be akin to ignoring the additive constant in an indefinite integral (Rindler 1969). In 1995, Wolfgang Priester of Bonn University found great significance in the late-time limit of the Friedmann-Lemaître equation, which defines the Hubble expansion rate as a function of redshift z:

$$H(z) = H_0\sqrt{\Omega_{M,0}(1+z)^3 + \Omega_{\Lambda,0} - (\Omega_{tot,0} - 1)(1+z)^2} \,. \qquad (11.1)$$

The matter ($\Omega_{M,0}$) term dominates in this equation at early times (high redshifts), while the vacuum one ($\Omega_{\Lambda,0}$) dominates at late ones. In fact, in the limit $t \to \infty$ it may be shown using the definition (7.7) that Eq. (11.1) becomes

$$\Lambda c^2 = 3H_\infty^2 \,, \qquad (11.2)$$

where H_∞ is the limiting value of the Hubble parameter as $t \to \infty$, assuming that this quantity exists (i.e., that the universe does not recollapse in the future). This equivalence between a fundamental constant of nature on the left and a dynamical quantity on the right suggested to Priester that setting $\Lambda = 0$ would amount to imposing an unreasonable degree of fine-tuning on the field equations of general

---

[1] This problem goes back to 1930 with Hubble's original and erroneously high value for the expansion rate (Kragh 1996, pp. 73–79 and references therein).

Fig. 11.1 The ten field equations of general relativity, with Λ restored to its rightful place (adapted by Wolfgang Priester from a woodcut by Gustav Doré; reproduced from Overduin et al. 2007)

relativity (Priester 1995), a viewpoint he pressed home with the sketch shown in Fig. 11.1. Put another way, if $\Lambda > 0$, then at sufficiently late times Eq. (11.2) immediately *predicts* that we will measure $\Omega_{\Lambda,0} = \Lambda c^2/3H_0^2 \sim 1$.

With the benefit of hindsight, we can now see that the discovery of dark energy did not spring completely out of nowhere in 1998, but was anticipated in several fascinating ways. One such "pre-discovery" was motivated primarily by inflation, and is sometimes referred to as the subtraction argument. Prior to the 1980s, cosmologists generally worked with one of two possible solutions of Einstein's equations, known as Friedmann-Robertson-Walker metrics. Those who placed the highest value on theoretical simplicity subscribed to the Einstein-de Sitter model ($\Omega_{M,0} = 1$, $\Omega_{\Lambda,0} = 0$). Indeed this became known as the "Standard Cold Dark Matter" (SCDM) model. Those who placed more trust in the growing empirical evidence for a low matter-density universe increasingly preferred a rival model known as "Open Cold Dark Matter" (OCDM) with $\Omega_{M,0} \approx 0.3$ and $\Omega_{\Lambda,0} = 0$. Neither model contained any vacuum energy. The successes of inflationary theory prompted some cosmologists to consider the possibility that the universe might have a low matter density and still be spatially flat, with $\Omega_{tot,0} = 1$. The difference between $\Omega_{tot,0}$ and $\Omega_{M,0}$ could only be made up by vacuum energy. This would later be called the "$\Lambda +$ Cold Dark Matter" or ΛCDM model. James Peebles at Princeton University presented an early case for such models in 1984, concluding that there was reasonable support for $\Omega_{M,0} \approx 0.2$ and $\Omega_{\Lambda,0} \approx 0.8$ (Peebles 1984).

Another pressing reason to consider a nonzero Λ-term was the lack of anisotropy observed in the cosmic microwave background (CMB) prior to 1992. This made it increasingly difficult to connect present-day large-scale structure to quantum fluctuations in the early universe—the crowning success of the inflationary paradigm. Already in 1933 Lemaître had argued that the loitering phase in a Λ-dominated universe would provide ideal conditions for galaxy formation (Kragh 1996, p. 53). This would ease the tension between the abundance of structure observed at low redshifts and the apparent lack of structure in the CMB. By the early 1990s, numerical simulations (Efstathiou et al. 1990; Martel and Wasserman 1990; Feldman and Evrard 1993) and analytic studies (Lahav et al. 1991; Sahni et al. 1992) showed that a significant Λ-term could indeed accelerate the growth of structure and reduce the required size of the initial perturbations. Conversely, a lower limit on the size of hoped-for CMB fluctuations was used to derive an *upper* limit of $\Omega_{\Lambda,0} < 0.7$ (Durrer and Straumann 1990). Some of these same studies, however, found that Λ had surprisingly little effect on measures of large-scale structure at late times, such as the distribution of galaxy velocities (Martel and Wasserman 1990) or the growth rate of density perturbations (Lahav et al. 1991). Particularly interesting in light of later developments was the proposal by Varun Sahni, Hume Feldman and Albert Stebbins at the Canadian Institute for Theoretical Astrophysics in Toronto to revive a closed loitering model of the Lemaître type (Sahni et al. 1992). Referring to the late phase of accelerated expansion that generically occurs in such models as vacuum energy begins to dominate over ordinary matter, they wrote:

> This period of rapid expansion may be called "late inflation" since the apparent horizon $\propto (\dot{R}/R)^{-1}$ grows less rapidly than the actual horizon ... *We propose that we may be living in the late inflationary phase*, although putting today in the loitering phase can also produce a viable mode [italics added].

This argument for dark energy became less pressing with the discovery of CMB anisotropies by the COsmic Background Explorer (COBE) satellite in 1992. However, numerical simulations of large-scale structure formation have continued to support the hypothesis that the present-day universe contains a significant fraction of vacuum energy (Fig. 11.2).

The first quantitative lower limits on $\Omega_{\Lambda,0}$ were set using galaxy number counts as a function of redshift. In principle, one should see more faint galaxies in a Λ-dominated universe than one with $\Lambda = 0$, because it contains more (comoving) volume at high redshifts. In practice, however, the number density of galaxies may change with redshift due to merging and galaxy formation, and the number we can observe may be affected by intrinsic evolution in galaxy luminosity. For these reasons, many cosmologists well into the 1990s concurred with the judgment of

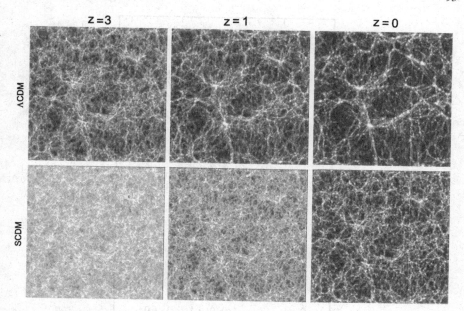

Fig. 11.2 Simulations of large-scale structure formation by the VIRGO Consortium (1998). In the *bottom row* is the Einstein-de Sitter model with no vacuum energy (labelled "SCDM" for "Standard Cold Dark Matter"). In the *top row* is the Λ-dominated *new* standard model (labelled CDM). Panel size is scaled to match the Hubble expansion; time runs from left (at redshift $z = 3$) to right (at $z = 0$). The Λ-dominated model provides a better fit to the observations. (*Image courtesy* J. Colberg and the VIRGO Consortium)

Allan Sandage at the Palomar Observatory, who had written in 1961 that "It would therefore seem impossible to find the correct world model from galaxy counts, and any attempt to do so appears to be a waste of telescope time" (Sandage 1961). In 1986, however, Edwin Loh at Princeton University made an initial attempt, obtaining $\Omega_{\Lambda,0} = 0.1^{+0.2}_{-0.4}$ for flat models under the assumption of constant galaxy number density (Loh 1986).

By 1988, observations indicated that galaxies would have had to merge at very high rates in a matter-only universe in order to bring the large number of faint galaxies at high redshifts down to the more modest population seen at $z = 0$. Using state-of-the-art galaxy evolution models, Masataka Fukugita at Kyoto University and his colleagues demonstrated in 1990 that the same observations could be fit without merging if $\Omega_{\Lambda,0} \approx 0.9$ (Fig. 11.3; Fukugita et al. 1990a). This work was heavily criticized at a time when theoretical prejudices among cosmologists still leaned heavily against a nonzero Λ-term.[2] It proved difficult to convincingly disentangle the effects of changes in number density from those in luminosity evolution (Gardner et al. 1993) and galaxy formation (Martel 1994). Attempts to account more fully

---

[2] Fukugita later recalled that invoking the cosmological constant was treated by many at the time as a "crime, or offense against the rules" (Fukugita 2014).

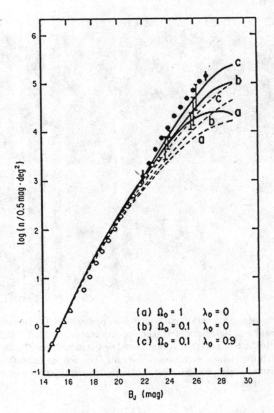

Fig. 11.3 Number counts of galaxies as a function of magnitude in the B_J (blue) band, from Fukugita et al. (1990a). The labeled curves correspond to predictions for flat models, as follows: **a** Einstein-de Sitter, **b** a variant of the OCDM model with $\Omega_{M,0}$ (here labeled as Ω_0) $= 0.1$, and **c** a ΛCDM-type model with $\Omega_{\Lambda,0}$ (here labeled λ_0) $= 0.9$. These curves incorporate a galaxy evolution model; the dashed lines show what happens when this evolution is neglected. (Reproduced by permission of M. Fukugita)

for evolutionary effects subsequently led to a limit of $\Omega_{\Lambda,0} > 0.53$ (Totani et al. 1997), ironically at a time when preliminary data from Type Ia supernova implied that $\Omega_{\Lambda,0} < 0.51$ (Perlmutter et al. 1997). The detection of dark energy by galaxy counts was not widely accepted until after similar lower limits had been obtained using larger supernovae samples, at which point the two methods were in rough agreement: $\Omega_{\Lambda,0} \approx 0.8$ (Totani and Yoshii 2000).

Quantitative *upper* limits on $\Omega_{\Lambda,0}$ were set during the 1990s using gravitational lenses. The increase in path length corresponding to a given redshift in vacuum-dominated models means that there are more sources to be lensed, and more lensed objects to be seen relative to matter-dominated models (Fukugita et al. 1990b). The first constraints to be derived in this way reinforced the majority opinion against a cosmological constant (Turner 1990):

Available data on the frequency of multiple image lensing of high-redshift quasars by galaxies suggest that *the cosmological constant cannot make a dominant contribution to producing a flat universe.*

In an article titled "Is there a cosmological constant?" Christopher Kochanek of the Harvard-Smithsonian Center for Astrophysics obtained a formal limit of $\Omega_{\Lambda,0} <$ 0.66 for flat models (Kochanek 1996). It was noted that this limit would be weakened if it should turn out that dust hides distant sources (Malhotra et al. 1997). But radio lenses would be far less affected, and these gave only slightly weaker constraints: $\Omega_{\Lambda,0} < 0.73$ for flat models (Falco et al. 1998). This upper limit was only marginally consistent with detection of dark energy that same year using Type Ia supernovae. The tension between the two ways of measuring Λ was put down to uncertainties in lens modeling (Chiba and Yoshii 1999) and calibration effects (Keeton 2002). By 2002, the two methods were in good agreement, with the supernova value having come down slightly and a larger sample of radio lenses giving a best-fit value of $\Omega_{\Lambda,0} = 0.69^{+0.14}_{-0.27}$ for flat models (Chae et al. 2002). At a symposium on gravitational lensing and cosmology in 2004, Dan Maoz of Tel-Aviv University asked (Maoz 2004):

This chain of events naturally raises the question of what went wrong ... Why did lensing statistics fail to predict the accelerating Universe before it was discovered by other means?

The conclusion was that a combination of both the above factors with extinction was likely responsible. But even though the agreement was belated, the fact that the existence of dark energy had now been confirmed by a completely independent technique was important.

Another way of measuring Λ was advocated by Wolfgang Priester and his colleagues beginning in 1991. The spectra of distant quasars show characteristic absorption lines making up what is known as the Lyman-α forest. These lines arise when the light from the quasars passes through intervening concentrations of gas which are thought to be distributed in sheets and filaments around emptier regions known as voids. Priester and his student Josef Hoell realized that these lines might be used as tracers of cosmic expansion, if intrinsic evolution in the absorber population could be neglected in comparison with the Hubble expansion rate. Their initial analysis (Hoell and Priester 1991) suggested that

If the Hubble constant is greater than $50 \, \text{km s}^{-1} \, \text{Mpc}^{-1}$ ... the cosmological constant must be greater than zero ... The apparently missing fraction of 90% or more of cosmic matter is entirely compensated for by the cosmological constant.

Together with Dierck-Ekkehard Liebscher of Potsdam University, Priester and Hoell applied this method to a set of high-resolution spectra from 21 quasars between 1992 and 1994 and found that the spacings of the absorption lines varied with redshift in a way that, if attributed entirely to cosmic expansion, implied an accelerating Λ-dominated world model that was *spatially closed.* Their best-fit model, which they termed the "Bonn-Potsdam" or BN-P model, was characterized by a dark-energy density of $\Omega_{\Lambda,0} \approx 1.08$ and a total matter density of $\Omega_{M,0} \approx 0.014$ (Liebscher and Priester 1992; Hoell et al. 1994).

Little attention was paid to the BN-P model. Its high value of $\Omega_{\Lambda,0}$ conflicted with the upper limits from lensing statistics, and its low value of $\Omega_{M,0}$ barely accommodated lower limits on the density of ordinary baryonic matter from big-bang nucleosynthesis, leaving no room for the dark matter. Its spatial curvature was at odds with the inflationary preference for exactly flat models. It also suffered from a form of "reverse age problem": its calculated age, 30 Gyr or more, greatly exceeded that of the oldest stars or galaxies yet observed, raising questions of why we should be living in an unusually young corner of the universe. The Achilles' heel of this method lay in its assumption that evolution in the absorber population could be neglected in comparison to the Hubble expansion rate. As can be seen in Fig. 11.4, this region of cosmological phase space is indeed picked out by data on Lyman-α absorbers—if their space density $n(z) \propto (1+z)^\eta$ is assumed not to evolve significantly with redshift z (i.e., they have an evolution parameter $\eta = 0$). However, there is reason to think that the comoving number density of absorbers should increase with redshift, since simulations show that the "walls" around voids at high redshift evolve gradually into filaments, and then into isolated clumps at low redshift (Overduin and Priester 2006). Figure 11.4b is an enlargment of the rectangular region in the upper left corner of Fig. 11.4a. It shows that an evolution parameter in the range $1.8 \lesssim \eta \lesssim 1.9$ rather than $\eta = 0$ would have led to a model close to CDM, rather than BN-P. The reasoning behind this method of measuring Λ, like that based on galaxy counts, was actually similar to that used five years later in the successful detection of dark energy with Type Ia supernovae. However, the systematic uncertainties associated with Lyman-α absorbers (and for that matter, with faint galaxies and lensed quasars) were too large to make them convincing probes of something as strange and unexpected as dark energy.

Nevertheless, by 1995 the stage was set. That year saw the publication of two articles, one by Jeremiah Ostriker and Paul Steinhardt whose original title (in preprint form) was "Cosmic concordance" (Ostriker and Steinhardt 1995) and the other by Lawrence Krauss and Michael Turner titled "The cosmological constant is back" (Krauss and Turner 1995). Both papers arrived at preferred models with $\Omega_{\Lambda,0} = 0.6 - 0.7$, focusing on the inflation-based subtraction argument but also laying new stress on the age problem, which was becoming particularly acute with increasingly compelling evidence for a "high value" of the Hubble constant greater than $70 \text{ km s}^{-1} \text{ Mpc}^{-1}$. Krauss and Turner wrote:

> Most important in this analysis is the fact that merely violating one of the constraints is not sufficient to allow a zero value of the cosmological constant. *Unless at least two of the fundamental observations described here are incorrect a cosmological constant is required by the data.*

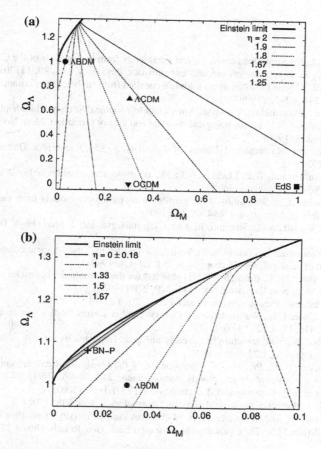

Fig. 11.4 a Best-fit values of the parameter η governing the evolution of the Lyman-α forest in the phase space defined by $\Omega_{M,0}$ and $\Omega_{\Lambda,0}$. Model universes above the solid curve marked "Einstein limit" do not originate in a big bang. **b** Enlargement of the rectangular region at the upper left corner of Fig. 11.4a. Inside the shaded region, the observational data on dn/dz are satisfied with no evolution at all in the Lyman-α forest. The labelled points correspond to world models discussed in the main text: EdS ($\Omega_{M,0} = 1$, $\Omega_{\Lambda,0} = 0$), OCDM (0.3,0), ΛCDM (0.3,0.7), Λ + baryonic matter only, or ΛBDM (0.03,1) and BN-P (0.014,1.08)

Ostriker and Steinhardt concluded their paper as follows:

> ... perhaps we have already identified models which, in broad outline, capture the essential properties of the large-scale universe. Should this be the case, a challenge arises: how can we explain the non-zero value of the cosmological constant from a theoretical point of view?

This was indeed the challenge, and it is still with us today.

References

Chae, K.-H., et al.: Constraints on cosmological parameters from the analysis of the Cosmic Lens All-Sky Survey radio-selected ravitational lens statistics. Phys. Rev. Lett. **89**, 151301, 4 (2002)

Chiba, M., Yoshii, Y.: New limits on a cosmological constant from statistics of gravitational lensing. Astrophys. J. **510**, 42–53 (1999)

Dicke, R.H.: Gravitation and the Universe. American Philosophical Society, Philadelphia (1970)

Durrer, R., Straumann, N.: The cosmological constant and galaxy formation. Mon. Not. R. Astron. Soc. **242**, 221–223 (1990)

Eddington, A.S.: The Mathematical Theory of Relativity, p. 154. Cambridge University Press, Cambridge (1924)

Efstathiou, G., Sutherland, W.J., Maddox, S.J.: The cosmological constant and cold dark matter. Nature **348**, 705–707 (1990)

Falco, E.E., Kochanek, C.S., Muñoz, J.A.: Limits on cosmological models from radio-selected gravitational lenses. Astrophys. J. **494**, 47–59 (1998)

Feldman, H.A., Evrard, A.E.: Structure in a loitering universe. Int. J. Mod. Phys. **D2**, 113–122 (1993)

Fukugita, M., Yamashita, K., Takahara, F., Yoshii, Y.: Test for the cosmological constant with the number count of faint galaxies. Astrophys. J. **361**, L1–L4 (1990a)

Fukugita, M., Futamase, T., Kasai, M.: A possible test for the cosmological constant with gravitational lenses. Mon. Not. R. Astron. Soc. **246**, 24P–27P (1990b)

Fukugita, M.: Personal communication. Email, Feb. 7, 2014 (2014)

Gardner, J.P., Cowie, L.L., Wainscoat, R.J.: Galaxy number counts from $K = 10$ to $K = 23$. Astrophys. J. **415**, L9–L12 (1993)

Hoell, J., Priester, W.: Void-structure in the early universe. Implications for a $\Lambda > 0$ cosmology. Astron. Astrophys. **251**, L23–L26 (1991)

Hoell, J., Liebscher, D.-E., Priester, W.: Confirmation of the Friedmann-Lemaître universe by the distribution of the larger absorbing clouds. Astron. Nachr. **315**, 89–96 (1994)

Keeton, C.R.: Rethinking lensing and Λ. Astrophys. J. **575**, L1–L4 (2002)

Kochanek, C.S.: Is there a cosmological constant? Astrophys. J. **466**, 638 (1996)

Kragh, H.: Cosmology and Controversy, pp. 52–53. Princeton University Press, Princeton (1996)

Krauss, L.M., Turner, M.S.: The cosmological constant is back. Gen. Relativ. Gravit. **27**, 1137–1144 (1995)

Lahav, O., Lilje, P.B., Primack, J.R., Rees, M.J.: Dynamical effects of the cosmological constant. Mon. Not. R. Astron. Soc. **251**, 128–136 (1991)

Liebscher, D.-E., Priester, W.: A new method to test the model of the universe. Astron. Astrophys. **261**, 377–381 (1992)

Loh, E.D.: Implications of the red-shift-number test for cosmology. Phys. Rev. Lett. **57**, 2865–2867 (1986)

Malhotra, S., Rhoads, J.E., Turner, E.L.: Through a lens darkly: evidence for dusty gravitational lenses. Mon. Not. R. Astron. Soc. **368**, 138–144 (1997)

Maoz, D.: Quasar lensing statistics and Ω_Λ: what went wrong? Proc. Int. Astron. Union **2004**, 413–418 (2004)

Martel, H., Wasserman, I.: Simulation of cosmological voids in $\Lambda > 0$ Friedmann models. Astrophys. J. **348**, 1–25 (1990)

Martel, H.: Galaxy formation in $\Lambda > 0$ Friedmann models: consequences for the number counts versus redshift test. Astrophys. J. **421**, L67–L70 (1994)

Ostriker, J.P., Steinhardt, P.J.: The observational case for a low-density universe with a non-zero cosmological constant. Nature **377**, 600–602 (1995). This paper was originally titled "Cosmic concordance" when it appeared in preprint form as arXiv:astro-ph/9505066

Overduin, J., Priester, W.: Quasar absorption-line number density in a closed, Λ-dominated universe. Astrophys. Space Sci. **305**, 159–163 (2006)

Overduin, J., Blome, H.-J., Hoell, J.: Wolfgang Priester: from the big bounce to the Λ-dominated universe. Naturwissenschaften **94**, 417–429 (2007)

Peebles, P.J.E.: Tests of cosmological models constrained by inflation. Astrophys. J. **284**, 439–444 (1984)

Peebles, P.J.E.: Principles of Physical Cosmology. Princeton University Press, Princeton (1993)

Perlmutter, S., et al.: Measurements of the cosmological parameters Ω and Λ from the first seven supernovae at $z \geq 0.35$. Astrophys. J. **483**, 565–581 (1997)

Priester, W.: The scale of the universe: a unit of length. Comments Astrophys. **17**, 327–342 (1995)

Rindler, W.: Essential Relativity: Special, General and Cosmological. Van Nostrand Reinhold, New York (1969)

Sahni, V., Feldman, H., Stebbins, A.: Loitering universe. Astrophys. J. **385**, 1–8 (1992)

Sandage, A.: The ability of the 200-inch telescope to discriminate between selected world models. Astrophys. J. **133**, 355–392 (1961)

Totani, T., Yoshii, Y., Sato, K.: Evolution of the luminosity density in the universe: implications for the nonzero cosmological constant. Astrophys. J. **483**, L75–L78 (1997)

Totani, T., Yoshii, Y.: Unavoidable selection effects in the analysis of faint galaxies in the Hubble Deep Field: probing the cosmology and merger history of galaxies. Astrophys. J. **540**, 81–98 (2000)

Turner, E.L.: Gravitational lensing limits on the cosmological constant in a flat universe. Astrophys. J. **365**, L43–L46 (1990)

Chapter 12
The Accelerating Universe

Abstract Improvements in detector technology, image analysis and other factors led to the definitive detection of dark energy in 1998 using the magnitude-redshift relation for Type Ia supernovae. However, the results were so shocking that even their own discoverers had trouble believing them at first. The rapid convergence of two independent and very different teams to a similar conclusion played a large part in convincing the cosmological community of the reality of a vacuum-dominated accelerating universe. But the deep theoretical issues regarding the status of the cosmological constant have only become murkier. We do not know why Λ is so much smaller than calculations based on quantum field theory would suggest; nor why the present energy densities of dark energy and matter should be so similar. In some senses, it appears that we are returning to a pre-Copernican view of the world, in which our composition and location in time are both special, even if our position in space is not.

Keywords Supernovae · Cosmic microwave background · Cosmological-constant problem · Coincidence problem

A decisive measurement of the density of dark energy was finally made using super-novae, whose intrinsic brightness allows them to be seen at great distances, and whose *consistency* in brightness makes them the first truly adequate standard candles for cosmology. Barring extinction, a supernova that is twice as far away as another will be four times fainter. Their difference in redshift can then be used to measure the extent to which space itself has stretched between them. Because vacuum energy inflates comoving distances at high redshift, it will "push" supernovae at those redshifts to greater physical distances, further dimming their apparent magnitudes relative to supernovae in more nearby galaxies.

At first, however, it was far from clear that supernovae could be observed in large enough numbers, and with small enough uncertainties, for a practical determination of what all expected would surely turn out to be an upper limit on Λ (Filippenko 2001). The first such long-term effort, by a Danish-led team, established the method

H. S. Kragh and J. M. Overduin, *The Weight of the Vacuum*, SpringerBriefs in Physics, DOI: 10.1007/978-3-642-55090-4_12, © The Author(s) 2014

that would later become standard: surveying patches of sky near successive new moons and looking for changes in brightness that might signal the appearance of a supernova, with rapid follow-up on promising candidates (Hansen et al. 1989; Norgaard-Nielsen et al. 1989). It turned up one new supernova, putting weak constraints on the deceleration parameter q_0, and hence on a combination of the present matter density $\Omega_{M,0}$ and dark-energy density $\Omega_{\Lambda,0}$ via $q_0 = \frac{1}{2}\Omega_{M,0} + \Omega_{\Lambda,0}$. Further breakthroughs had to await two developments: technological progress in large-format detectors and data analysis, and a better understanding of the systematics of supernova explosions themselves.

During this time, the Supernova Cosmology Project (SCP) formed at the Lawrence Berkeley Laboratory in California with the express aim of measuring q_0 to higher precision using Type Ia supernovae or SNe Ia (Perlmutter et al. 1988, Pennypacker et al. 1989). These objects, which result from the explosions of hydrogen-poor white dwarf stars in binary systems, promised both the brightest peak luminosity and greatest degree of relative homogeneity. Still, their variation in peak brightness was significantly larger than the expected size of the dimming effect in, say, the Open Cold Dark Matter (OCDM) model relative to the then still-dominant Einstein-de Sitter (EdS) model. Nevertheless, with refined image-processing and analysis techniques, the SCP was able to use the discovery of a high-redshift supernova in 1992 to show that a low-q_0 (OCDM) model was favoured over the flat EdS model with $q_0 = 0.5$ (Perlmutter et al. 1995). In a sign of the continuing strength of theoretical prejudices, the cosmological constant was not mentioned in this paper, which assumed $\Lambda = 0$. By 1994 the number of high-redshift supernovae discovered by the SCP had risen to seven, leading to the first tentative constraints on $\Omega_{\Lambda,0}$ as well as $\Omega_{M,0}$ (Perlmutter et al. 1997). In marked contrast with later developments, it was found that $\Omega_{\Lambda,0} < 0.51$ at the 95 % confidence level, with a preferred value of $\Omega_{\Lambda,0} = 0.06^{+0.28}_{-0.34}$. The authors stated that:

> The results for Ω_Λ-versus-Ω_M are inconsistent with Λ-dominated, low-density, flat cosmologies that have been proposed to reconcile the ages of globular cluster stars with higher Hubble constant values.

In 1993, Mark Phillips at the Cerro Tololo Inter-American Observatory in Chile showed that the variation in peak luminosity among nearby SNe Ia was closely correlated with their time taken to reach peak brightness, allowing the high-redshift supernovae to be much more precise calibrated (Phillips 1993). Motivated by this discovery, and by the growing success of the SCP in finding supernovae, a second group known as the High-Z supernova search Team (HZT) joined the hunt in 1994. The fact that there were now two independent and philosophically quite different teams of detectives on the trail of the same quarry would prove very valuable. Alex Filippenko, the only person to work closely with both teams, later recalled (Filippenko 2001):

> The SCP ... included some astronomers, but its clear center of activity was LBL [the Lawrence Berkeley Lab in California], and it consisted largely of people trained within the high-energy physics community ... Like the SCP, the HZT was an international team— but in contrast to the SCP, it consisted primarily of astronomers ... [and] ... was structured in a less hierarchical manner.

The HZT discovered its first high-redshift supernova in 1995, resulting in a preliminary constraint quite different from that published by the SCP: $\Omega_{\Lambda,0} = 0.6^{+0.4}_{-0.5}$ for spatially flat models (Schmidt et al. 1998).

By 1997 the HZT had discovered a total of 16 high-z SNe Ia, and the SCP had amassed a collection of 42 (with only two in common). Both groups were beginning to realize that the evidence pointed to an accelerating universe dominated by dark energy, a possibility which they discussed—at first tentatively—at scientific meetings beginning in 1998. The HZT was the first to publish its conclusion that $\Omega_{\Lambda,0} = 0.76 \pm 0.10$ in a flat universe (Riess et al. 1998). The SCP followed soon after with results expressed in the form (Perlmutter et al. 1999):

$$0.8\Omega_{M,0} - 0.6\Omega_{\Lambda,0} \approx -0.2 \pm 0.1. \tag{12.1}$$

For a spatially flat universe with $\Omega_{M,0} = 1 - \Omega_{\Lambda,0}$, this implied $\Omega_{\Lambda,0} \approx 0.71 \pm 0.07$. The authors summarized their new findings as follows:

> The data are strongly inconsistent with the $\Lambda = 0$, flat universe model ... that has been the theoretically favored cosmology. If the simplest inflationary theories are correct and the universe is spatially flat, then the supernova data imply that there is a significant, positive cosmological constant. ... [But] even if the universe is not flat, the confidence regions ... suggest that the cosmological constant is a significant constituent of the energy density of the universe.

Such was the combined weight of evidence from both teams that, already by May of 1998, a two-thirds majority of theoretical physicists at a hastily convened meeting in Chicago indicated that they were now ready to accept the reality of a nonzero cosmological constant (Filippenko 2001).

The subsequent discovery of many more supernovae has only bolstered the case for a dark energy-dominated universe. Figure 12.1 plots the apparent magnitudes of about 200 SNe Ia compiled by Tonry et al. (2003) and Riess et al. (2007) as a function of redshift, relative to what they would be in a ΛCDM model. (This difference is referred to as the "magnitude residual.") Theoretical residuals corresponding to some of the standard cosmological models are shown and labeled for comparison. Redshifts $z \gtrsim 1.5$ are particularly important, because they probe the transition regime between the present, low-redshift universe (accelerating and dominated by dark energy) and the high-redshift one (decelerating and dominated by ordinary matter). A steadily increasing degree of dimming with redshift might plausibly be attributed to systematic effects, such as evolution or extinction by "grey" dust—but it would be highly implausible for such effects to turn off or reverse themselves at exactly this epoch. Indeed, the data are of sufficient quantity and quality that they can be used to set constraints, not only on models with a cosmological constant, but also those with a time-varying cosmological term (Overduin et al. 2007). A recent compilation and analysis of 580 supernovae essentially confirms the original HZT and SCP results from 14 years earlier, finding that for flat models (Suzuki et al. 2012, 95% confidence):

$$\Omega_{\Lambda,0} = 0.73 \pm 0.03. \tag{12.2}$$

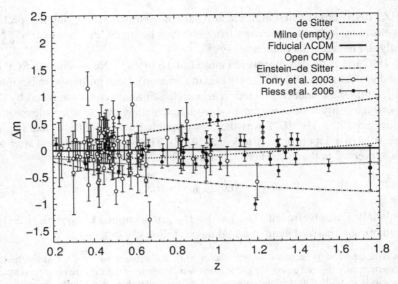

Fig. 12.1 Magnitude-redshift diagram showing magnitude residuals (relative to the ΛCDM model) for about 200 Type Ia supernova observed by Tonry et al. (2003) and Riess et al. (2007), together with theoretical predictions for some of the standard cosmological models

Further support for the existence of dark energy has arisen from a completely independent source: the angular power spectrum of CMB fluctuations. These are produced by density waves in the primordial plasma, "the oldest music in the universe" (Lineweaver 2001). The first peak in the power spectrum picks out the angular size of the largest fluctuations in this plasma at the moment when the universe became transparent to light. Because it is seen through the "lens" of a curved universe, the location of this peak is sensitive to the total density $\Omega_{tot,0} = \Omega_{\Lambda,0} + \Omega_{M,0}$. Constraints on this parameter were first obtained from the balloon-borne experiments MAXIMA (Balbi et al. 2000) and BOOMERanG (Lange et al. 2001), and combined with data from the COBE satellite to give $\Omega_{tot,0} = 1.11^{+0.13}_{-0.12}$ at 95 % confidence (Jaffe et al. 2001). This measurement is impervious to most of the uncertainties discussed so far, because it leapfrogs "local" systems whose interpretation is complex (supernovae, galaxies, and quasars) and goes directly back to the radiation-dominated era, when physics was simpler.

Because the CMB and SNe Ia constraints are nearly orthogonal in the $\Omega_{M,0} - \Omega_{\Lambda,0}$ plane (Fig. 12.2), the combination of both kinds of data together puts strong limits on the density of dark energy alone: $\Omega_{\Lambda,0} = 0.75 \pm 0.14$ at 95 % confidence (Jaffe et al. 2001). Moreover, the fact that the favoured regions picked out by the CMB and SNe data intersect at $\Omega_{M,0} \approx 0.3$ agrees with decades of independent evidence from clusters of galaxies (whose masses can be inferred from gravitational lensing and hot X-ray emitting gas). This in turn serves as a vindication of the the general

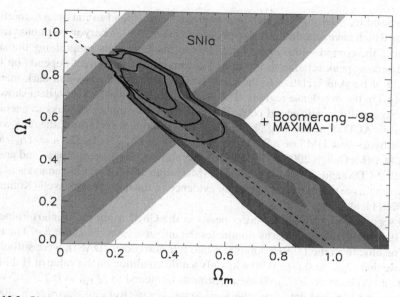

Fig. 12.2 Observational constraints on the values of $\Omega_{M,0}$ and $\Omega_{\Lambda,0}$ from both SNe and CMB observations (COBE, MAXIMA, BOOMERanG). Shown are 68, 95 and 99.7 % confidence intervals inferred both separately and jointly from the data (Reprinted from Jaffe et al. 2001 by permission of Jaffe and Richard)

picture of large-scale structure formation via gravitational instability (although the required dark matter has not yet been identified).

The Wilkinson Microwave Anisotropy Probe (WMAP), a successor to COBE, has reinforced these results, finding $\Omega_{tot,0} = 1.02 \pm 0.02$ at 95 % confidence (Bennett et al. 2003). The Λ-dominated universe is therefore spatially flat, or very close to it. The location of the first peak in these data actually hints at "marginally closed" models, although the implied departure from flatness is not statistically significant and could be explained in other ways (White et al. 2000). If spatial flatness is assumed, then the WMAP data are best fit by a ΛCDM-type model with (Bennett et al. 2003, 95 % confidence):

$$\Omega_{\Lambda,0} = 0.73 \pm 0.04. \tag{12.3}$$

in agreement with Eq. (12.2) from SNeIa. The same data impose a limit on the equation-of-state parameter of $w < -0.78$ (Bennett et al. 2003), consistent with the hypothesis that dark energy is a pure cosmological constant ($w = -1$). A repeat of the BOOMERanG experiment has further tightened this latter result to $w = -0.86 \pm 0.36$ (from the CMB alone), or $w = -0.93^{+0.093}_{-0.097}$ (when combined with supernova data; MacTavish et al. 2006).

Much attention is focused on the higher-order peaks of the CMB power spectrum, which contain valuable clues about the matter component. Odd-numbered peaks are produced by regions of the primordial plasma which have been maximally compressed by infalling material, and even ones correspond to maximally rarefied regions which have rebounded due to photon pressure. A high baryon-to-photon ratio enhances the compressions and retards the rarefractions, thus suppressing the size of the second peak relative to the first. The strength of this effect depends on the fraction of baryons (relative to the more weakly bound neutrinos and dark-matter particles) in the overdense regions. The BOOMERanG and MAXIMA data showed an unexpectedly small second peak. While there are a number of ways to account for this in ΛCDM models (e.g., by "tilting" the primordial spectrum), the data were also fit by a "no-CDM" or ΛBDM-type model with $\Omega_{cdm} = 0$, $\Omega_{M,0} = \Omega_{bar}$ and $\Omega_{\Lambda,0} \approx 1$ (McGaugh 2000). The WMAP data show a larger second peak and are fit by both ΛCDM and ΛBDM-type models (McGaugh 2004). Ongoing analysis of the data from this experiment has turned up evidence of the third peak as well (Komatsu et al. 2011).

Observation of even higher-order peaks in the CMB using radio interferometry has established that dark energy dominates the universe *even if it is not flat*. The first such results, from the Degree Angular Scale Interferometer (DASI), were sufficient to prove that $\Omega_{\Lambda,0} = 0.60 \pm 0.30$ with only a mild condition on the value of Hubble's constant (Pryke et al. 2002). These constraints tightened to $\Omega_{\Lambda,0} = 0.72^{+0.14}_{-0.26}$ with analysis of additional data from the Very Small Array (Rubiño-Martin et al. 2003) and $\Omega_{\Lambda,0} = 0.67^{+0.20}_{-0.26}$ from the Cosmic Background Imager (CBI; Sievers et al. 2003). With the results from a fourth experiment, the Atacama Cosmology Telescope (ACT), even the requirement of a prior on the Hubble parameter is dropped, allowing a convincing detection of dark energy from CMB fluctuations alone: $\Omega_{\Lambda,0} = 0.61^{+0.23}_{-0.29}$ (95 % confidence; Sherwin et al. 2011). This is possible because the resolution of the higher-order peaks is now high enough to reveal the effects of gravitational lensing by large-scale structure, breaking a geometric degeneracy that weakened previous limits. The first release of data from the Planck satellite has improved the precision of this CMB-only limit to $\Omega_{\Lambda,0} = 0.69 \pm 0.04$ (Ade et al. 2013), consistent with the measurement by WMAP assuming flatness, Eq. (12.3). Other ways of measuring the density of dark energy continue to be devised, including X-ray observations of the hot gas in massive, dynamically relaxed galaxy clusters (Allen et al. 2004) and the detection of "baryon acoustic oscillations," echoes of sound waves in the primordial plasma that are imprinted onto the present-day distribution of galaxies (Eisenstein et al. 2005). All continue to agree with the basic ΛCDM model with $\Omega_{\Lambda,0} \approx 0.7$ (Mehta et al. 2012, Mantz et al. 2014).

The depth of the change in thinking that has been triggered by these developments on the observational side can hardly be exaggerated. Only twenty years ago, it was routine to set $\Lambda = 0$ and cosmologists had two main choices: the "one true faith" (flat, with $\Omega_{M,0} \equiv 1$), or the "reformed" (open, with individual believers being free to choose their own values near $\Omega_{M,0} \approx 0.3$). All this has been irrevocably altered by the SNe and CMB experiments. If there is a guiding principle now, it is no longer $\Omega_{M,0} \approx 0.3$, and certainly not $\Omega_{\Lambda,0} = 0$; it is $\Omega_{tot,0} \approx 1$ from the power spectrum of the CMB. Cosmologists have been obliged to accept a Λ-term, and it is not a question of whether or not it dominates the energy budget of the universe, but by *how much*. The resulting picture is self-consistent, and agrees with nearly all the

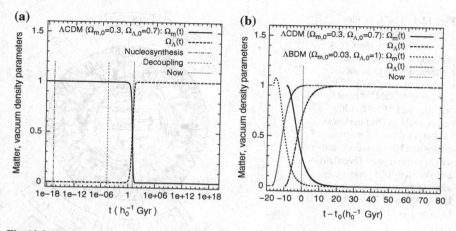

Fig. 12.3 The evolution of $\Omega_M(t)$ and $\Omega_\Lambda(t)$ in vacuum-dominated models. Panel **a** shows a single model (ΛCDM) over twenty powers of time in either direction. Plotted this way, we are seen to live at a very special time (marked "Now"). Standard nucleosynthesis ($t_{nuc} \sim 1$ s) and matter-radiation decoupling times ($t_{dec} \sim 10^{11}$ s) are included for comparison. Panel **b** shows both the ΛCDM and ΛBDM models on a linear rather than logarithmic scale, for the first $100\,h_0^{-1}$ Gyr after the big bang (i.e. the lifetime of the stars and galaxies)

data. Major questions, however, remain on the theoretical side. Prime among these is the problem of the cosmological constant (Chap. 10), which becomes all the more acute now that we definitely know that $\Lambda > 0$. Previously, one might have hoped that a new symmetry of nature, akin to supersymmetry, might force Λ to vanish (Weinberg 1989). Such an explanation is not tenable if Λ exists but at levels 120 orders of magnitude below theoretical expectations.

A related concern has to do with the *evolution* of the matter and dark-energy density parameters Ω_M and Ω_Λ over time, given by

$$\Omega_M(t) \equiv \frac{\rho_M(t)}{\rho_{crit}(t)} = \frac{\Omega_{M,0}}{\tilde{R}^3(t)\,\tilde{H}^2(t)}\,, \qquad \Omega_\Lambda(t) \equiv \frac{\rho_\Lambda(t)}{\rho_{crit}(t)} = \frac{\Omega_{\Lambda,0}}{\tilde{H}^2(t)}\,, \qquad (12.4)$$

where $\tilde{R}(t) \equiv R(t)/R(t_0)$, $\tilde{H}(t) \equiv H(t)/H_0$, and t_0 is the present time. These equations can be solved exactly for spatially flat models. At early times, dark energy is insignificant relative to matter ($\Omega_\Lambda \sim 0$) and ($\Omega_M \sim 1$), but the situation is reversed at late times when $\Omega_\Lambda \sim 1$ and $\Omega_M \sim 0$. These statements hold regardless of the presently measured densities of matter or dark energy. Results for the ΛCDM model are illustrated in Fig. 12.3a.

What is remarkable in this figure is the location of the present in relation to the values of Ω_M and Ω_Λ. We have apparently arrived on the scene at the precise moment when these two quantities are in the midst of changing places. This is the same coincidence problem that was originally cited by Robert Dicke as the chief

Fig. 12.4 The four elements of ancient cosmology (*top*, attributed to the Greek philosopher Empedocles), contrasted with their modern counterparts (*bottom*). The subscripts "CDM," "BAR," "LUM," and "BDM" refer respectively to cold dark matter, baryonic matter, luminous matter and baryonic dark matter (Reprinted from Overduin and Priester 2001; originally adapted from a figure in a 1519 edition of Aristotle's *Libri de Caelo*)

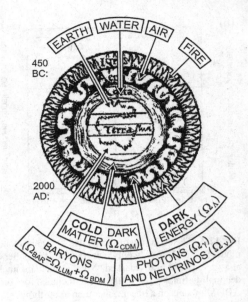

objection to dark energy in general. Sean Carroll (2001) has aptly described such a universe as "preposterous," writing:

> This scenario staggers under the burden of its unnaturalness, but nevertheless crosses the finish line well ahead of any of its competitors by agreeing so well with the data.

Cosmology may indeed be moving toward a position like that of particle physics, where a standard model accounts for all observed phenomena to high precision, but is founded on a set of fundamental parameters so finely-tuned as to suggest that the underlying reality has not yet been grasped.

In Fig. 12.3b the same evolution is plotted on a *linear*, rather than logarithmic scale in time over approximately the lifetime of the galaxies (as determined by their main-sequence stellar populations). One would not, after all, expect observers to appear on the scene long after all the galaxies had disappeared, or in the early stages of the expanding fireball. Seen from the perspective of Fig. 12.3b, the coincidence, while still striking, is perhaps no longer so preposterous. The ΛCDM model still appears fine-tuned, in that "Now" follows rather quickly on the heels of the epoch of matter-vacuum equality. In the ΛBDM model, $\Omega_{M,0}$ and $\Omega_{\Lambda,0}$ are closer to the cosmological time-averages of $\Omega_M(t)$ and $\Omega_\Lambda(t)$ (namely zero and one respectively). In such a picture it might be easier to believe that we have not arrived on the scene at a special time, but merely a *late* one.

In either case, however, the universe of modern cosmology hardly resembles the one we see. It is composed to a first approximation of invisible dark energy whose physical origin remains obscure. Most of what remains is in the form of cold dark matter particles, whose identity is also unknown. Close inspection is needed to make out the further contribution of neutrinos, although this too is nonzero. And baryons, the stuff of which we are made, are little more than a cosmic afterthought.

This picture entails a shift in the way we see the universe that is profound enough to be called a Copernican counter-revolution. For while our location in space may be undistinguished, our composition, as well as the time in which we live, appear to be truly atypical. The new four elements are shown together with their ancient counterparts in Fig. 12.4.

References

Ade, P.A.R., et al.: Planck 2013 results. XVI. Cosmological parameters. Astron. Astrophys. in press (2013) arXiv:1303.5076 [astro-ph]

Allen, S.W., et al.: Constraints on dark energy from Chandra observations of the largest relaxed galaxy clusters. Mon. Not. R. Astron. Soc. 353, 457–467 (2004)

Balbi, A., et al.: Constraints on cosmological parameters from MAXIMA-1. Astrophys. J. 545, L1–L4 (2000)

Bennett, C.L., et al.: First year Wilkinson Microwave Anisotropy Probe (WMAP) observations: preliminary maps and basic results. Astrophys. J. Suppl. Ser. 148, 1–27 (2003)

Carroll, S.M.: The cosmological constant. Living Rev. Relativ. 4, 1 (2001)

Eisenstein, D.J., et al.: Detection of the baryon acoustic peak in the large-scale correlation function of SDSS luminous red galaxies. Astrophys. J. 633, 560574 (2005)

Filippenko, A.V.: Einstein's biggest blunder? High-redshift supernovae and the accelerating universe. Publ. Astron. Soc. Pac. 113, 1441–1448 (2001)

Hansen, L., et al.: A supernova at $z = 0.28$ and the rate of distant supernovae. Astron. Astrophys. 211, L9–L11 (1989)

Jaffe, A.H., et al.: Cosmology from MAXIMA-1, BOOMERANG, and COBE DMR cosmic microwave background observations. Phys. Rev. Lett. 86, 3475 (2001)

Komatsu, E., et al.: Seven-year Wilkinson Microwave Anistropy Probe (WMAP) observations: cosmological interpretation. Astrophys. J. 192, 18, 47 (2011)

Lange, A.E., et al.: Cosmological parameters from the first results of Boomerang. Phys. Rev. D 63, 042001 (2001)

Lineweaver, C.H.: CMBology. In:. Brainerd T.G, Kochanek C.S. (eds.) Gravitational Lensing: Recent Progress and Future Goals, Astronomical Society of the Pacific, Vol. 237, pp. 111–120, San Francisco (2001)

MacTavish, C.J., et al.: Cosmological parameters from the 2003 flight of Boomerang. Astrophys. J. 647, 799–812 (2006)

Mantz, A.B., et al.: Cosmology and astrophysics from relaxed galaxy clusters II: cosmological constraints. Mont. Not. R. Astron. Soc. 440, 2077–2098 (2014) arXiv:1402.6212 [astro-ph]

McGaugh, S.S.: Boomerang data suggest a purely baryonic universe. Astrophys. J. 541, L33–L36 (2000)

McGaugh, S.S.: Confrontation of modified Newtonian dynamics predictions with Wilkinson Microwave Anisotropy Probe first year data. Astrophys. J. 611, 26–39 (2004)

Mehta, K.T., et al.: A 2 % distance to $z = 0.35$ by reconstructing baryon acoustic oscillations— III: cosmological measurements and interpretation. Mon. Not. R. Astron. Soc. 427, 2168–2179 (2012)

Norgaard-Nielsen, H.U., et al.: The discovery of a type Ia supernova at a redshift of 0.31. Nature 339, 523–525 (1989)

Overduin, J., Priester, W.: Problems of modern cosmology: how dominant is the vacuum? Naturwissenschaften 88, 229–248 (2001)

Overduin, J.M., Wesson, P.S., Mashhoon, B.: Decaying dark energy in higher-dimensional gravity. Astron. Astrophys. 473, 727–731 (2007)

Pennypacker, C., et al.: Automated supernova discoveries: status of the Berkeley project. In E.B. Norman (ed.) Particle Astrophysics: Forefront Experimental Issues. Proceedings of a Workshop in Berkeley, California, pp. 188–189. World Scientific, Singapore, 8–10 Dec 1988 (1989)

Perlmutter, S., et al.: The status of Berkeley's realtime supernova search. In: Proceedings of the Ninth Santa Cruz Summer Workshop in Astronomy and Astrophysics, Springer, New York, 13–24 July 1987 (1988)

Perlmutter, S., et al.: A supernova at $z = 0.458$ and implications for measuring the cosmological deceleration. Astrophys. J. **440**, L41–L44 (1995)

Perlmutter, S., et al.: Measurements of the cosmological parameters Ω and Λ from the first seven supernovae at $z \geq 0.35$. Astrophys. J. **483**, 565–581 (1997)

Perlmutter, S., et al.: Measurements of Ω and Λ from 42 high-redshift supernovae. Astrophys. J. **517**, 565–586 (1999)

Phillips, M.M.: The absolute magnitudes of type Ia supernovae. Astrophys. J. **413**, L105–L108 (1993)

Pryke, C., et al.: Cosmological parameter extraction from the first season of observations with the Degree-scale Angular Scale Interferometer. Astrophys. J. **568**, 46–51 (2002)

Riess, A.G., et al.: Observational evidence from supernovae for an accelerating universe and a cosmological constant. Astrophys. J. **116**, 1009–1038 (1998)

Riess, A.G., et al.: New Hubble Space Telescope discoveries of type Ia supernovae at $z > 1$: narrowing constraints on the early behavior of dark energy. Asrtophys. J. **659**, 98–121 (2007)

Rubiño-Martin, J.A., et al.: First results from the Very Small Array— IV. Cosmological parameter estimation. Mon. Not. R. Astron. Soc. **341**, 1084–1092 (2003)

Schmidt, B.P., et al.: The high-Z supernova search: measuring cosmic deceleration and global curvature of the universe using type Ia supernovae. Astrophys. J. **507**, 46–63 (1998)

Sherwin, B.D., et al.: Evidence for dark energy from the cosmic microwave background alone using the Atacama Cosmology Telescope lensing measurements. Phys. Rev. Lett. **107**, 021302 (2011)

Sievers, J.L., et al.: Cosmological parameters from Cosmic Background Imager observations and comparisons with Boomerang, DASI and MAXIMA. Astrophys. J. **591**, 599–622 (2003)

Suzuki, N., et al.: The Hubble Space Telescope cluster supernova survey. V. Improving the dark-energy constraints above z 1 and building an early-type-hosted supernova sample. Astrophys. J. **746**, 85, 24 (2012)

Tonry, J.L., et al.: Cosmological results from high-z supernovae. Astrophys. J. **594**, 1–24 (2003)

Weinberg, S.: The cosmological constant problem. Rev, Mod. Phys. **61**, 1–23 (1989)

White, M., Scott, D., Pierpaoli, E.: Boomerang returns unexpectedly. Astrophys. J. **545**, 1–5 (2000)

Author Index

H. S. Kragh and J. M. Overduin, *The Weight of the Vacuum*, SpringerBriefs in Physics,
DOI: 10.1007/978-3-642-55090-4, © The Author(s) 2014